# HIGH-RESOLUTION MASS SPECTROSCOPY FOR PHYTOCHEMICAL ANALYSIS

*State-of-the-Art Applications and Techniques*

# HIGH-RESOLUTION MASS SPECTROSCOPY FOR PHYTOCHEMICAL ANALYSIS

*State-of-the-Art Applications and Techniques*

*Edited by:*

**Sreeraj Gopi, PhD**
**Augustine Amalraj, PhD**
**Shintu Jude, MSc**

First edition published 2022

**Apple Academic Press Inc.**
1265 Goldenrod Circle, NE,
Palm Bay, FL 32905 USA

4164 Lakeshore Road, Burlington,
ON, L7L 1A4 Canada

**CRC Press**
6000 Broken Sound Parkway NW,
Suite 300, Boca Raton, FL 33487-2742 USA

2 Park Square, Milton Park,
Abingdon, Oxon, OX14 4RN UK

**Library and Archives Canada Cataloguing in Publication**

Title: High-resolution mass spectroscopy for phytochemical analysis : state-of-the-art applications and techniques / edited by: Sreeraj Gopi, PhD, Augustine Amalraj, PhD, Shintu Jude, MSc.

Names: Gopi, Sreeraj, editor. | Amalraj, Augustine, editor. | Jude, Shintu, editor.

Description: First edition. | Includes bibliographical references and index.

Identifiers: Canadiana (print) 2021016316X | Canadiana (ebook) 20210163283 | ISBN 9781771889964 (hardcover) | ISBN 9781774638187 (softcover) | ISBN 9781003153146 (ebook)

Subjects: LCSH: Phytochemicals—Analysis. | LCSH: High resolution spectroscopy.

Classification: LCC QK865 .H54 2022 | DDC 572/.2—dc23

**Library of Congress Cataloging-in-Publication Data**

Names: Gopi, Sreeraj, editor. | Amalraj, Augustine, editor. | Jude, Shintu, editor.

Title: High-resolution mass spectroscopy for phytochemical analysis : state-of-the-art applications and techniques / Sreeraj Gopi, Augustine Amalraj, Shintu Jude.

Description: First edition. | Palm Bay, FL, USA : Apple Academic Press, 2022. | Includes bibliographical references and index. | Summary: "This new volume provides a bird's-eye view of the properties, utilization, and importance of high resolution mass spectrometry (HRMS) for phytochemical analyses. The book discusses the new and state-of-the-art technologies related to HRMS in phytochemical analysis for the food industry in a comprehensive manner. Phytochemical characterization of plants is important in the food and nutraceutical industries and is also necessary in the procedures followed for drug development, toxicology determination, forensic studies, origin verification, quality assurance, etc. Easy determination of active compounds and isolation as well as purification of the same from natural matrices are required, and the possibilities and advantages of HRMS pave the way for improved analysis patterns in phytochemistry. This book is unique in that its sole consideration is on the importance of HRMS in the field of phytochemical analysis. Along with an overview of basic instrumental information, the volume provides a detailed account of data processing and dereplication strategies. Technologies such as bioanalytical techniques and bioassays are considered also to provide support for the functions of the instruments used. In addition, a case study is presented to depict the complete phytochemical characterization of a matrix by HRMS. The book covers processing and computational techniques, dereplication, hyphenation, high-resolution bioassays, bioanalytical screening/purification techniques, applications of gas chromatography-high-resolution mass spectrometry, and more. Key features: Covers the fundamental instrumentation and techniques Discusses HRMS-based phytochemical research details Focuses strictly on the phytochemical considerations High-Resolution Mass Spectroscopy for Phytochemical Analysis: State-of-the-Art Applications and Techniques will be a valuable reference guide and resource for researchers, faculty and students in related fields, as well as those in the phytochemical industries"-- Provided by publisher.

Identifiers: LCCN 2021011439 (print) | LCCN 2021011440 (ebook) | ISBN 9781771889964 (hardback) | ISBN 9781774638187 (paperback) | ISBN 9781003153146 (ebook)

Subjects: LCSH: Phytochemicals--Analysis. | High resolution spectroscopy.

Classification: LCC QK865 .H585 2022 (print) | LCC QK865 (ebook) | DDC 572/.2--dc23

LC record available at https://lccn.loc.gov/2021011439

LC ebook record available at https://lccn.loc.gov/2021011440

ISBN: 978-1-77188-996-4 (hbk)
ISBN: 978-1-77463-818-7 (pbk)
ISBN: 978-1-00315-314-6 (ebk)

# About the Editors

**Sreeraj Gopi, PhD**
*Plant Lipids Private Limited, Kerala, India*

Sreeraj Gopi, PhD, is an Industrial Scientist with a doctorate in organic chemistry, nanotechnology, and nanodrug delivery. He has been working in the area of natural products, isolation, and establishing biological activities. He has published more than 85 international articles and filed more than 75 international patents. Dr. Gopi is a fellow of the Royal Society of Chemistry.

**Augustine Amalraj, PhD**
*Deputy Manager, Department of Research and Development,*
*Plant Lipids Private Limited, Cochin, India*

Augustine Amalraj, PhD, is currently working as a Deputy Manager in the Department of Research and Development in Plant Lipids Private Limited, Cochin, India. He obtained his doctoral degree in chemistry from Gandhigram Rural Institute-Deemed University, Gandhigram, Tamil Nadu, India. His research interest is in applied chemistry, food chemistry, natural product chemistry, environmental chemistry, chemosensors, and polymer and nanocomposite materials. He has published more than 50 research articles in international journals as well as 10 book chapters.

**Shintu Jude, MSc**
*Assistant Manager, Research and Development Department,*
*Plant Lipids Private Limited, Cochin, India*

Shintu Jude is working as an Assistant Manager in the Research and Development Department at Plant Lipids Private Limited, Cochin, India. Mrs. Jude completed her postgraduation at Mahatma Gandhi University. She is working on mass spectroscopic instruments for natural products and metabolites. She has published more than 20 research articles and 10 book chapters.

# Contents

# Contributors

**Janet Adeyinka Adebiyi**
Department of Biotechnology and Food Technology, Faculty of Science,
University of Johannesburg, P.O. Box – 17011, Doornfontein, Johannesburg,
South Africa, E-mail: janetaadex@gmail.com

**Oluwafemi Ayodeji Adebo**
Department of Biotechnology and Food Technology, Faculty of Science,
University of Johannesburg, P.O. Box – 17011, Doornfontein, Johannesburg, South Africa,
E-mail: oadebo@uj.ac.za

**Gbenga Adedeji Adewumi**
Department of Microbiology, Faculty of Science, University of Lagos, Akoka, Lagos, Nigeria

**Rosa M. Alonso-Salces**
Research Group of Applied Microbiology, Social Bee Research Center, Institute for Research in
Production, Health and Environment, CONICET, Department of Biology,
Faculty of Exact and Natural Sciences, National University of Mar del Plata, Funes – 3350,
Mar del Plata – 7600, Argentina

**Luis A. Berrueta**
Department of Analytical Chemistry, Faculty of Science and Technology,
University of the Basque Country/Euskal HerrikoUnibertsitatea (UPV/EHU),
PO Box – 644, 48080, Bilbao, Spain

**Blanca Gallo**
Department of Analytical Chemistry, Faculty of Science and Technology,
University of the Basque Country/Euskal HerrikoUnibertsitatea (UPV/EHU),
PO Box – 644, 48080, Bilbao, Spain

**Sreeraj Gopi**
Research and Development (R&D) Center, Plant Lipids (P) Ltd., Kadayiruppu,
Kolenchery, Cochin, Ernakulam, Kerala – 682311, India

**Joby Jacob**
Research and Development (R&D) Center, Plant Lipids (P) Ltd., Kadayiruppu,
Kolenchery, Cochin, Ernakulam, Kerala – 682311, India

**Shintu Jude**
Research and Development (R&D) Center, Plant Lipids (P) Ltd., Kadayiruppu,
Kolenchery, Cochin, Ernakulam, Kerala – 682311, India

**Anjana S. Nair**
Research and Development (R&D) Center, Plant Lipids (P) Ltd., Kadayiruppu,
Kolenchery, Cochin, Ernakulam, Kerala – 682311, India

**Nomali Ngobese**
Department of Botany and Plant Biotechnology, Faculty of Science, University of Johannesburg,
P.O. Box – 524, Auckland Park, Johannesburg, South Africa

**Patrick Berka Njobeh**
Department of Biotechnology and Food Technology, Faculty of Science,
University of Johannesburg, P.O. Box – 17011, Doornfontein, Johannesburg, South Africa

**Gabriela E. Viacava**
Research Group of Food Engineering, National Council of Scientific and Technological Research
(CONICET), Department of Chemistry and Food Engineering, Faculty of Engineering,
National University of Mar del Plata, 4302 Juan B. Justo Street, Mar del Plata – 7600, Argentina

# Abbreviations

| | |
|---|---|
| 2D | two-dimensional |
| ACE | angiotensin-converting enzyme |
| AChE | acetylcholinesterase |
| AD | Alzheimer's disease |
| AGH | alpha-glucosidase |
| AGN-TCMB | AGH magnetic nanoparticle beads |
| AIF | all ion fragmentation |
| AMDIS | automated mass spectral deconvolution and identification system |
| AMPK | adenosine monophosphate-activated protein kinase |
| AMT | accurate mass and time |
| ANOVA | analysis of variance |
| APCI | atmospheric pressure chemical ionization |
| APGC-MS | atmospheric pressure gas chromatography equipped with tandem mass spectrometry |
| APPI | atmospheric pressure photoionization |
| ARE | antioxidant response element |
| ASAP | atmospheric pressure solids analysis probe |
| CAD | collisionally activated dissociation |
| CAP-e | cell-based antioxidant protection in erythrocytes |
| CAS | chemical abstracts service |
| CCIC | Campus Chemical Instrument Center |
| CE | capillary electrophoresis |
| CEM | channel electron multiplier |
| CF | chemical fingerprinting |
| CGAs | chlorogenic acids |
| CGE | capillary gel electrophoresis |
| CI | chemical ionization |
| CID | collision-induced dissociation |
| CIEF | capillary isoelectric focusing |
| CMA | caffeoylmalic acid |

| | |
|---|---|
| CMC | cell membrane chromatography |
| COX | cyclooxygenase |
| CPs | chlorinated paraffin |
| CQA | caffeoylquinic acid |
| CVDHD | cardiovascular disease herbal database |
| CysLTs | cysteinyl leukotrienes |
| CZE | capillary zone electrophoresis |
| Da | Daltons |
| DAD | diode array detector |
| DAPCI | desorption atmospheric pressure chemical ionization |
| DAPPI | desorption atmospheric pressure photoionization |
| DART | direct analysis in real time |
| DCBI | desorption corona beam ionization |
| DESI | desorption electrospray ionization |
| DIA | data-independent acquisition |
| DL-PCBs | dioxin-like polychlorinated biphenyls |
| DOX | doxorubicin |
| DP | degrees of polymerization |
| DP | dynamic programming |
| ECD | electronic circular dichroism |
| ECG | epicatechin gallate |
| ECNI | electron capture negative ionization |
| EESI | extractive electrospray ionization |
| EGCG | epigallocatechin gallate |
| EGFR | epithelial cell growth factor receptor |
| EI | electron ionization |
| EIC | extracted ion chromatogram |
| ELDI | electrospray-assisted desorption/ionization |
| ESI | electrospray ionization |
| FAAH | fatty acid amide hydrolase |
| FAB | fast-atom bombardment |
| FAC | frontal affinity chromatography |
| FBF | find by formula |
| FD | field desorption |
| FI | field ionization |
| FIA | flow injection analysis |

| | |
|---|---|
| FPD | focal plane detectors |
| FT | Fourier transform |
| FT-ICR | Fourier transform ion cyclotron resonance |
| FWHM | full width at half maximum |
| GC-EI-Orbitrap-HRAMS | gas chromatography-electron ionization-Orbitrap-high resolution accurate mass spectrometry |
| GCG | gallocatechin gallate |
| GC-HRAMS | gas chromatography-high resolution accurate mass spectrometry |
| GC-HR-MS | gas chromatography-high-resolution mass spectrometry |
| GC-HRToF-MS | gas chromatography-high resolution time of flight mass spectrometry |
| GC-MS | gas chromatography-mass spectrometry |
| G-LC | gas-liquid chromatography |
| G-SC | gas-solid chromatography |
| GSL | glucosinolates |
| HCA | hierarchical clustering analysis |
| HRFSMS | high-resolution full scan mass spectrometric |
| HRMS | high-resolution mass spectrometers |
| HSCCC | high-speed counter-current chromatography |
| HSM | hydrophobic subtraction model |
| IC/PAD | ion chromatography coupled with pulsed amperometric detection |
| ICPs | instant coffee premixes |
| ICR | ion cyclotron resonance |
| ID-GC-HRMS | isotope dilution high-resolution gas chromatography/high-resolution mass spectrometry |
| IMERs | immobilized enzyme reactors |
| IT | ion trap |
| JCMPIH | Jatropha curcas meal protein isolate hydrolysates |
| LAB | lactic acid bacteria |
| LADESI | laser-assisted desorption electrospray ionization |
| LAESI | laser ablation electrospray ionization |
| LC-MS | liquid chromatography-mass spectrometry |

| | |
|---|---|
| LDI | laser desorption ionization |
| LOX | lipoxygenase |
| LRMS | low-resolution mass spectrometry |
| LSS | linear solvent strength |
| LT | leukotriene |
| LTQ | linear trap quadrupole |
| MAF | maximum autocorrelation factor |
| MALDESI | matrix-assisted laser desorption electrospray ionization |
| MALDI | matrix-assisted laser desorption ionization |
| MCCP | medium carbon chain CP |
| MCP | microchannel plates |
| MDS | multidimensional scaling |
| MEF | multistage elemental formula |
| MRM | multiple reactions monitoring |
| MS | mass spectrometer |
| MW | molecular weights |
| NACE | non-aqueous capillary electrophoresis |
| NIMS | nanoelectrospray ionization, nanostructure-initiator mass spectrometry |
| NINA | nontargeted diagnostic ion network analysis |
| NMF | non-negative matrix factorization |
| NMR | nuclear magnetic resonance |
| NNC | nearest neighboring connecting |
| NO | nitric oxide |
| NPACT | naturally occurring plant-based anticancerous compound-activity-target database |
| NPASS | natural product activity and species source |
| NPS | new psychoactive substances |
| NSDESI | nanospray desorption electrospray ionization |
| OPLC | overpressured layer chromatography |
| OPLS-DA | orthogonal projection to latent structures discriminate analysis |
| ORAC | oxygen radical absorbance capacity |
| PA | phenolic acids |
| PAHs | polyaromatic hydrocarbons |
| PAINS | pan-assay interference compounds |
| PAs | proanthocyanidins |

| PBDEs | polybrominated diphenyl ethers |
|-------|-------------------------------|
| PCA | principal component analysis |
| PCBs | polychlorinated biphenyls |
| PCO | Polygonum Cillinerve Ohwi |
| PDE5 | phosphodiesterase 5 |
| PFE | pressurized fluid extraction |
| PhGs | phenylethanoid glycosides |
| PIF | precursor ion fingerprinting |
| PlantMAT | plant metabolite annotation toolbox |
| PLE | pressurized liquid extraction |
| PLSA | probabilistic latent semantic analysis |
| PLS-DA | partial least squares discriminant analysis |
| PM | plasma membrane |
| POPs | persistent organic pollutants |
| PPARγ | peroxisome proliferator-activated receptor-γ |
| PTP1B | protein-tyrosine phosphatase 1B |
| pYES | planar yeast estrogen screen |
| QC | quality control |
| Q-Orbitrap | quadrupole Orbitrap |
| QqQ | triple quadrupole |
| QSRR | quantitative structure- retention relationship |
| QToF | quadrupole time-of-flight |
| QToF/MS | quadrupole-time-of-flightmass analyzer |
| RA | relative abundance |
| RF | radio frequency |
| RHS | roasted hazelnut skin |
| RIPs | reverse inhibitors of peroxidase |
| RP | reversed-phase |
| RP-HPLC | reverse-phase high-performance liquid chromatography |
| RPLC | reverse-phase liquid chromatography |
| RT | retention time |
| SA | similarity analysis |
| SAR | structure-activity relationship |
| SCCP | short carbon chain CP |
| SEC | size exclusion chromatography |
| SEM | secondary electron multipliers |
| SFC | supercritical fluid chromatography |

| | |
|---|---|
| SGA | steroidal glycoalkaloids |
| SIM | selected ion monitoring |
| SIMS | secondary ion mass spectroscopy |
| SORI-CID | sustained off-resonance irradiation collision-induced dissociation |
| SPE | solid-phase extraction |
| SZRD | Suanzaoren decoction |
| TCM | traditional Chinese medicine |
| TFC | total flavonoid content |
| TICs | total ion chromatograms |
| TLC | thin layer chromatography |
| ToF | time-of-flight |
| TPC | total phenolic content |
| TQ | triple quadrupole |
| UHPLC | ultrahigh performance liquid chromatography |
| UHPLC-DAD-HRAM-MS | ultra-high-performance liquid chromatography-diode array detection-high resolution accurate mass-mass spectrometry |
| XEQJ | Xiao-Er-Qing-Jie |
| XOD | xanthine oxidase |
| YCHD | Yinchenhao decoction |
| YXST | Yangxinshi tablet |

# Preface

The exact determination of molecular mass provides a strong path towards compound characterization. High-resolution mass spectrometry (HRMS) relies on the fact that an individual atom's mass is not just a factor of atomic mass units. In the case of molecules, the mere addition of atomic mass units of atoms doesn't deliver any meaningful idea. In addition, the different molecules of the same compound can be of different masses, as there is a possibility of occurrence of isotopes. Therefore, it is possible to confidently assume the exact molecular mass's elemental composition, thereby laying a foundation for many analytical procedures. Also, HRMS's better features in terms of sensitivity, selectivity, resolution, repeatability, etc., improve the quality of analysis and allow the distinction of compounds from even complex matrices like herbal products.

Plant-derived products are important in the industries for their application as food, drug, nutraceutical, color, flavor properties, and many more. Hence, phytochemical characterization is an inevitable term in many aspects. Phytochemical characterization and related techniques are included in the procedures followed for drug development, toxicology determination, forensic studies, origin verification, quality assurance, etc.

Plants serve as the precursors for many drugs in traditional as well as modern medicinal systems. Even synthetic drugs take a model from the herbal moieties. Even so, the chaos related to synthetic products always brightens the path of natural products. Easy determination of active compounds and isolation, as well as purification of the same from natural matrices, raise the bigger question here; and the answer comes from many technologies and techniques, with HRMS being the base of them. The book discusses the newer technologies related to HRMS in the phytochemical analysis.

Along with basic instrumental information, a detailed account is given on the data processing and dereplication strategies. The technologies such as bioanalytical techniques and bioassays are considered in a way such that they provide support for the functions of the instrument. A case study is presented to depict the complete phytochemical characterization of a matrix by HRMS. In addition, the relevance of HRMS in the field of the food industry is discussed in a comprehensive manner. Altogether, the book provides a bird's-eye view of the properties and utilization of HRMS for phytochemical analyzes.

# CHAPTER 1

# High-Resolution Mass Spectrometry: Instrumentation in General

SHINTU JUDE and SREERAJ GOPI

*Research and Development (R&D) Center, Plant Lipids (P) Ltd., Kadayiruppu, Kolenchery, Cochin, Ernakulam, Kerala – 682311, India*

## ABSTRACT

Mass spectrometry is becoming the nucleus of many analytical instruments and study protocols nowadays and has been revolutionized by technologies with more and more mass-resolving potential. Today's high-resolution mass spectrometry uses a variety of mass spectrometry designs. However, they are all still based on the simple, yet most important resolving power. This chapter introduces the fundamentals of high-resolution mass spectrometry-including detectors, ionization techniques, and data acquisition—in an approachable and comprehensive way. It goes beyond the basics of the instrumentation and provides a detailed look at the techniques and plays behind the instrumentation.

## 1.1 INTRODUCTION

Mass spectrometry holds a prime chair in the venue of analysis. Mass spectrometry deals with the measurement of mass to charge ratio (m/z) of the analytes of interest. Therefore, the procedure starts from sample introduction, proceeds with ionization into charged particles, and finally, detection of the same as depicted in Figure 1.1. There are many customized/ improved versions of these mentioned steps that are developed to improve the efficiency, sensitivity, simplicity, etc., of mass spectrometers (MSs). There is a huge number of studies presented on the potential of MSs.

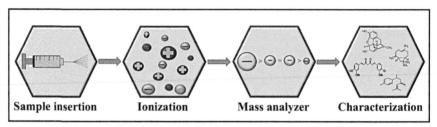

| Sample insertion | Ionization | Mass analyzer | Characterization |

**FIGURE 1.1** Schematic diagram representing the basic HRMS analysis.

The invention of MS was a milestone in the field of analysis. It made both the qualitative and quantitative analyzes more rapid, robust, and accurate by their immanent sensitivity and selectivity. As technology gets advanced, the drawbacks also find their new face to show off. Of course, mass analyzers are the unsung heroes in targeted quantitation. At the same time, due to their targeted nature, they may not identify the new/reformed compounds in the samples, though they are present in considerable amounts. Normally these quantitative analyzes are done in unit resolution, which can cause interference from compounds of similar mass and can lead to an alteration in the observed results. In addition, for the complex samples containing many analytes in it, method development takes much time, as the transitions should be optimized for each and every analyte separately. This scenario was altered by the evolution of high-resolution mass spectrometers (HRMS) in the 1960s with the launch of double-focusing magnetic-sector mass instruments [1].

## 1.2 HIGH-RESOLUTION MASS SPECTROMETERS (HRMS)

The resolving power of an instrument can be defined as the ability of the instrument to provide well separation between neighboring peaks. The term resolution bears a little different meaning. It is defined and expressed as the ratio of the mass of interest to the difference in mass, which is the width of a peak at a particular peak height. High resolution or the sufficiently resolved peaks allows accurate mass determination by the instrumentation, which is proved to be applicable for the qualitative analysis, including elemental composition [2]. The improvements in terms of resolution, sensitivity, and speed enable it for the quantitative analysis also.

In Figure 1.2, the dotted lines represent the highly resolved peaks. The resolution of a mass peak can be indicated as Eqn. (1):

$$R = M/\Delta M \tag{1}$$

where, M is the mass of the analyte and $\Delta M$ is peak width/ peak separation. If the point of consideration is the peak width, then it can be better mentioned by the term 'full width at half maximum (FWHM),' which denotes the width of a spectrum curve at half of its maximum amplitude (Figure 1.2(B)). As can be seen in the figure and equation, the lower the FWHM better will be the resolution.

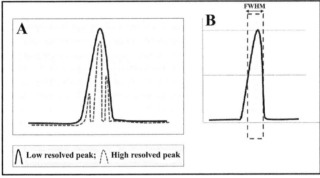

**FIGURE 1.2** (A) Difference between highly resolved peak and low resolved one; (B) representation of FWHM.

## 1.3 INSTRUMENTATION

The HRMS family includes Fourier transform ion cyclotron resonance (FT-ICR), time-of-flight (ToF), and Orbitrap mass analyzers. This generation of high-resolution techniques could directly identify the molecular formula of the compound of interest by a single injection. Thus HRMS detectors are "mass" detectives in the field of chemical investigation since they help to find out the answers to the questions-what, why, how, when, and where. Advance in science found that hybrid HRMS instruments those constructed by combining different MSs in a single instrument such as quadrupole ToF (Q-ToF), ion trap (IT)-ToF, linear trap quadrupole (LTQ)-Orbitrap, or Qe-Orbitrap, can act more intelligently. However, the basic instrumentation setup remains the same and can be represented as in Figure 1.1 irrespective of its resolution, and the basics of major high-resolution techniques are discussed in detail in this chapter.

### 1.3.1   TIME OF FLIGHT (TOF)

Chromatography, equipped with the time of flight analyzers, was the first endeavor of commercial high-resolution MS, and it was indicated as a mature technology [3]. As mentioned by the developers of ToF instrumentation, it is a "velocitron," which separates the ions with respect to velocity [4]. After the sample introduction, the analyte moieties are subjected to ionization, in order to produce charged ion packets, which are then directed by using an ion pulser to move along a defined path inside the flight tube (ToF tube). In the defined conditions of voltage and acceleration, the ions tend to move depending on their mass, and the ions with lighter mass to charge ratio (m/z) value will move faster and reach the detector first. The detector measures both the flight time and number of arriving ions simultaneously. Thus it enables the measurement of both the m/z value and its abundance at the same time. The working of ToF is governed by the ToF relationship, which can be expressed as in Eqn. (2):

$$m = (2E/d^2) \times t^2 \tag{2}$$

where; $m$ is the detected mass (m/z ratio), $E$ is the energy to which an ion is accelerated, $d$ is the flight path distance, and $t$ is the flight time.

Two variations of ToF systems are introduced to the scene, according to the direction of the acceleration of ions: linear and orthogonal. As the latter allows efficient coupling of continuous ion sources and provides better resolving power, it is used commonly for various applications. Figure 1.3 represents the working of orthogonal acceleration ToF-MS. Advanced ToF systems have many add-ons such as space focusing (to reduce the distance between produced ions in the ion source), reflectron (ion mirror-to compensate the initial energy difference between the produced ions, and to increase the flight path) [5], analog to digital conversion detector (ADC-to record multiple ion events), ToF/ToF configurations, etc. However, in ToF systems, the ion separation depends majorly on the effective flight tube length. In addition, they may be affected by saturation, which creates a negative impact on the mass detection, identification, and signal strength. In cases of extremely high masses, the resolution and related mass accuracy get deteriorated [6]. Moreover, in the case of ultra-small molecule mass ranges, they may provide false identification [7].

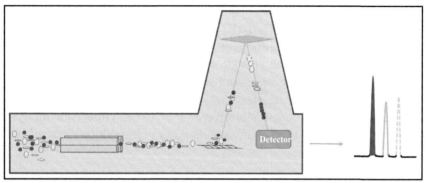

**FIGURE 1.3**    Schematic presentation of orthogonal TOF-MS.

## *1.3.2    FOURIER TRANSFORM ION CYCLOTRON RESONANCE (FT-ICR)*

FT-ICR was developed to determine the m/z value of ions accurately by an entirely different technology, devoid of the drawbacks of ToF [8]. Here, the produced ions are collected in a system of a strong magnetic field consisting of electric trapping plates, which is known as a penning trap. An oscillating electric field-commonly radio frequency (RF) pulses-is applied through the excitation plates for a while, orthogonal in direction to the magnetic field. As the cyclotron frequency appears as a function of mass, each ion possesses a distinct cyclotron frequency, and the ions which have cyclotron frequency equal to the emitted frequency will be excited to produce rotating ion packets. The rotation speed and frequency are dependent on the mass of the ions. The excited ions continue to rotate in their resonant cyclotron frequencies, with an increasing orbit, even after removing the excitation field. These ions attain a maximum orbit radius so as to come closer to the electrodes pair in the detection plates and induce a charge on them. The ions will lose energy, reduces their radius, and regain their original orbit. The induced charges on the plates will be detected as an image current, and the resulting free induction decay signals are then submitted to discrete fast Fourier transformation (Figure 1.4). Different ions possess different ion cyclotron frequencies depending on their mass, and the Fourier transformation signals give a spectrum of cyclotron frequencies corresponding to the m/z values. During the analysis, these frequencies and respective image current resulting from the charged ions are plotted over a period of time, which contains both the ion frequencies and their intensities corresponding to the abundance. This time-domain signal will be converted into a mass spectrum. FT-ICR is a technology duo, in which Fourier transform (FT) denotes the representation of masses as frequencies, and ICR mentions it's mode of working. The excitation of ions to larger trajectories to provide sufficient angular frequency is called ICR. Frequencies can be measured with high accuracy, and so, FT-MS can provide highly accurate mass measurements and high-resolution mass spectra. The angular frequency of a moving ion of mass m and charge q, in uniform magnetic field B, can be calculated as in Eqn. (3):

$$\omega = qB/m \tag{3}$$

In addition, FT-ICR makes use of this basic cyclotron equation. A schematic diagram of the FT-ICR is given in Figure 1.4.

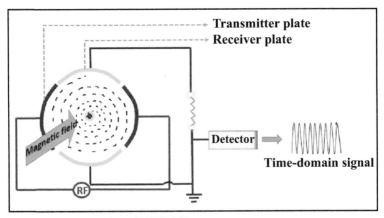

**FIGURE 1.4**   Representation of FTICRMS.

### 1.3.3   ORBITRAP

Orbitrap mass analyzers are the new faces in the field, and as the name mentions, they work by orbital trapping of ions. The instrumentation is an advanced form of Kingdon ion trap, which traps the charged ions in an electrostatic field and separates the ions by applying a voltage between the two coaxial electrodes and thereby making the ions cycle around the inner electrode [9]. Orbitrap modified its predecessor by taking the advanced ideologies of FT mass analyzers. It consists of two electrodes arranged coaxially—the spindle-shaped central electrode and the outer barrier electrode, which actually consists of two parts arranged face to face and separated by a narrow dielectric gap. One part of the outer electrode functions as an ion exciter, and the other part as a detector. The void between the coaxial electrodes is the measurement chamber. A voltage is applied between the coaxial electrodes to create an electrostatic field in the direction of the axis. The ions are collected in a cooled curved trap (known as C-trap), injected at once through the ion entrance in the outer electrode, and are attracted towards the inner electrode due to the electric field. However, the tangential velocity of ions compensates for this attraction force. These two different forces make the ions circulate in the inner electrode forming a "rotating ion ring" and oscillate along the axis at the same time. Ions with different m/z value form different rotating rings which oscillate with different frequencies. The frequency of the axial oscillations is free from the initial properties of the ion, except the m/z ratio, and can be used for the mass measurement. The axial oscillation frequency is detected by the outer detector electrode and produces a corresponding image current, which is then converted to a chromatogram corresponding to the mass to charge spectrum as in the case of FTICR. The axial frequency of the ions can be expressed by Eqn. (4):

$$\omega = \sqrt{[k/(m/z)]} \tag{4}$$

where; $\omega$ is the oscillation frequency, and $k$ is the instrument constant [10]. The working of orbitrap is depicted in Figure 1.5.

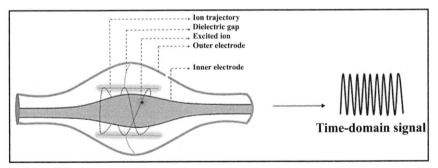

**FIGURE 1.5**  Orbitrap.

In many cases, orbitrap exhibits more sensitivity and mass resolving capacity than the other two aforementioned techniques. However, this excellence loses gradually with the increase in m/z value, the number of ions, and the spectral acquisition rate. This drawback affects more while hyphenating the technique with faster separation methods. In such cases, the performance of orbitrap drops, especially for chemical species with higher m/z values. Moreover, poor fidelity of isotope pattern, narrow dynamic range, etc., are also reported for orbitrap [11].

## 1.4   SAMPLE INTRODUCTION

The introduction of the sample in a proper manner is the first consideration in almost all analytical techniques. The mode of sample introduction/ inlet depends on the nature of the analyte and the sample matrix. Mass spectrometry basically relies on the generation of ions and hence on the ionizability of molecules. Thermally stable volatile compounds, which can exert a high vapor pressure at least by heating, are able to introduce into the system directly as a gas phase through a direct vapor inlet, which is renowned as the simplest sample introduction method. While coupled with separation techniques, the differentiated compounds are entered into the mass instrument through an interface. TLC separated samples can be evaluated in mass instruments either by direct introduction or by further preparation, depending on the mode of ionization applied and the instruments in the hyphenation series. In the case of GC, the interface is designed such that not any carrier gas, but all the analytes enter into the source region. Liquid chromatography allows a number of interfaces with a potential for ionization from the condensed phase, which enables the analyzes of thermally labile compounds also. Ionization probe is another major technique for the introduction of samples. Here, the low vapor pressure samples can be placed directly in the source region by means of a probe. Frontal elution paper chromatography utilizes paper for mounting the samples. ESI techniques can be supported by the usage of the needle, wooden tip, Al foil, etc., as the sample introduction media. In ambient ionization techniques like DESI, DART, etc., the sample itself can be produced for ionization without any preparation, irrespective of its nature.

## 1.5   IONIZATION TECHNIQUES

During HRMS analyzes, the detectability and sensitivity are influenced by the employed mode and extend of ionization. So, maximizing ion generation is the first major hurdle to be jumped over to achieve high resolution and sensitivity. This has led to more profound research in the area, resulted in the proliferation of many more modern techniques. It may take another complete book to discuss all of them in detail. Therefore, consideration is given to the significant technologies, which allow HRMS to be a strong tool across the area of phytochemical analysis.

According to the mode and extend of ionization, the techniques fall under different categories. Gas-phase ionization comprises electron ionization (EI), chemical ionization (CI), fast-atom bombardment (FAB), etc., in which the compounds with thermal stability have been volatilized, and the gas phase molecules are ionized. Fast atom bombardment (FAB) is a first-generation ionization method, which makes use of a fast atom beam for the ionization. However, most of these classic techniques are associated with high fragmentation and are not considered for HRMS, as the nature of information needed from the samples is different. The most frequently used HRMS ionization techniques are electrospray ionization (ESI), secondary ion mass spectroscopy (SIMS), atmospheric pressure chemical ionization (APCI), atmospheric pressure photoionization (APPI), desorption electrospray ionization (DESI), and matrix-assisted laser desorption ionization (MALDI).

ESI is a widely used technology, which makes use of an electrospray, which is obtained by applying a definite voltage across the droplets from the capillary under atmospheric pressure. An inert de-solvating gas, together with high temperature, succeeds in the evaporation of the solvent, producing independent analyte ions. This technique is apt for highly polar, thermally labile, large molecules and can create multiply charged ions. Heated ESI is a little more advanced technology, where the heated nozzle improves electrostatic fields, thus enhances ionization, and increases the possibility to do the analyzes in positive or negative modes, depending on the nature and the pH of the target compound. Besides, it allows the provision of multiple ionization for a single molecule, which permits the analyzes of large molecules. However, apart from the possibilities, ESI constrains itself from using a range of solvents and thus closes the chances for many applications.

The presence of a corona discharge in the ionization region differentiates APCI from ESI. Corona discharge furnishes a high-density discharge current in the ionization area, which causes the excitation of nebulization gas, producing the molecular ions of the same. Besides, the high probe temperature and regulated nebulization gas flow together produce a gas stream from the analyte solution. Here, an ion-molecule reaction takes place between the molecular ions of nebulization gas and the evaporated mobile phase, which in turn ionizes the analyte molecule in the gas stream.

Like in APCI, APPI requires the analyte solution to get vaporized. It utilizes a UV light source for the primary ionization, and the analyte molecule absorbs a high-energy photon and subsequently ejects an electron to form a radical cation. Direct APPI produces ions directly from the analyte molecules, forming molecular radical cations. Dopant APPI employs photoionizable molecules (dopants) for creating charged species, and they undergo charge exchange reactions to produce analyte molecular ions. APPI can ionize less polar compounds, compared to APCI and ESI.

SIMS is considered the most sensitive surface analysis technique used widely for solid samples. A primary ion beam is allowed to bombard with the sample surface, in order to generate secondary ions, which is analyzed by a MS. While carrying out SIMS, the sample surface gets sputtered. It is possible to obtain the ion fragmentation patterns from an atomic monolayer of the surface, with a controlled sputtering rate. This is useful for molecular species identification, and the process is named 'static SIMS.' In 'dynamic SIMS,' a high yield of secondary ions is produced by a high sputtering rate, which provides information for the depth profile and quantitative data.

Matrix-assisted laser desorption/ionization (MALDI) is a potential device for soft ionization of large molecules like protein. Though the tool is using to a great degree, the mechanism behind MALDI is still under investigation. The sample to be ionized is mixed well with a matrix solution so that on drying, the analyte would be embedded within the recrystallized matrix. A beam of the laser is used to illuminate the sample, co-crystallized with the matrix, and placed in the vacuum. This illumination results in both desorption and ionization, producing molecular ions in the vapor phase. The invention of MALDI has increased the range of analyzable samples in terms of type, molecular mass, and functionalities, and in combining with mass instruments, it provides a strong platform for bioanalysis.

Literally, the term field desorption (FD) denotes the desorption of a material into its gas phase ions under a strong potential field from a metal surface (known as an emitter) on which it was deposited. The ionization can happen if the metal surface has an appropriate geometry under a high vacuum, and it can be thermal ionization or field ionization (FI).

Thermally labile compounds with high molecular mass are ionized well with FD. Ionization of the volatile/gaseous compounds by applying a high potential field is termed as FI. The mechanism provides molecular ions and is suitable for less polar and thermally stable compounds.

### 1.5.1   AMBIENT IONIZATION TECHNIQUES

An important innovation in the domain of mass spectrometry ionization techniques is ambient ionization, which allows the ionization of samples in their ambient environment outside the MS, without any treatment or prior preparation and within a little time. It utilizes the scope of selective desorption and ionization of compounds, which occurs at the surface. By employing direct ionization, the possibilities of complications associated with sample preparation and matrix effects are avoided. It is really interesting to know that more than thirty ambient ionization techniques were developed within the last two decades. These techniques, while hyphenated with MS, find applications in a number of fields. Among the ambient ionization techniques, DESI, direct analysis in real time (DART), desorption atmospheric pressure photoionization (DAPPI) are the most accepted and widely used.

DESI is a combo of desorption and ESI. Here, by using the electrospray, a fast-moving, electrically charged mist of ions and charged droplets were produced, and the same was directed towards the sample surface to carry out the ionization. Ionization occurs by charge transfer, and the secondary ions emitted from the sample surface are mass analyzed. DESI provides ionization for a huge range of masses and materials [12]. In DART, a heated stream of inert gas is excited by using glow discharge plasma and is directed to the sample surface to be analyzed. Ion-molecule reactions take place, and as a result, analyte ions are produced. As exhibited by DESI, DART also characterized by low energy ionization, a large range of sample surfaces, good sensitivities, and can be operated in open environments without sample preparation. These special natures of DART enable it to do wonders; even living organisms can be subjected to DART-MS [13]. DAPPI enables desorption of the analyte from the sample surface by employing a heated jet of vaporized solvent. The desorbed analyte molecules from the surface are then subjected to ionization induced by the photons emitted from the lamp. Comparing with other ambient ionization techniques, DAPPI is more sensitive towards less polar analytes. Highly conjugated compounds can be ionized selectively with less matrix contaminations. DAPCI resembles APCI, having corona discharge for generating species from chemicals, but proceeds with desorption or ionization of surface molecules. In most of the cases, ambient air or solvents serve as chemical reagents, and proceeds for small molecules with volatile nature [14].

The second generation of ionization techniques are introduced and utilized by combining more than one properties of them and was named accordingly, such as matrix-assisted laser desorption electrospray ionization (MALDESI), extractive electrospray ionization (EESI), laser ablation electrospray ionization (LAESI), laser-assisted desorption electrospray ionization (LADESI), electrospray-assisted desorption/ionization (ELDI), nanospray desorption electrospray ionization (NSDESI), desorption atmospheric pressure chemical ionization (DAPCI), desorption atmospheric pressure photoionization (DAPPI), chip-based nanoelectrospray ionization, nanostructure-initiator mass spectrometry (NIMS), and atmospheric pressure solids analysis probe (ASAP). The increased number of techniques allows one to select any of them according to the sample and the conditions. The advanced benefits such as sustained minimal usage of solvents allowing long runs of samples, minimum sample consumption, and reduced cross contaminations make these 2G ionization techniques very special [15].

## 1.6  DETECTORS

All the processes starting from the sample introduction and propagated through suitable ionization come to a fruitful end, when the ions resolved by the analyzer are detected by proper detectors. Basically, the detector identifies and measures the charged particle and provides data on the presence and abundance of the particular charged species. There are many types of detectors invented by the time, some of which are discussed below.

A metal cup, which is highly conductive and can catch the ions in a vacuum, can be employed as a basic mass detector, and this design is known as a Faraday cup. Here, the ions coming from the analyzer deposit their charge on the cup (electrode), which produces a corresponding electric current, flowing away from the electrode. It produces a voltage when passing through a resistor of high impedance.

As mass spectroscopy handles the small amount of samples, the number of particles passing through the detector will be quite small. Hence the detectors with amplification serve well. Secondary electron multipliers (SEM), discrete dynode electron multipliers, multiple plate detectors, etc., follows this consideration. Here, the ions with high velocity, coming from the analyzer, happen to hit on the surface of a metal/semiconductor, and a large number of secondary electrons are produced from the surface.

An electrode, keeping a more positive potential and kept opposite to the emission location, could attract these secondary electrons and causes the emission of more electrons each. This generates an electric current to be detected by a preamplifier. In some cases, the secondary electron production made cascades inside a long tube in order to multiply the gain and is termed as a channel electron multiplier (CEM). The length and diameter of the tube determine the gain. Thus, the size of these CEM tubes is reduced up to micrometers and arranged into a bundle to form microchannel plates (MCP), which allows the detection of all the ions at a particular time interval (Figure 1.6).

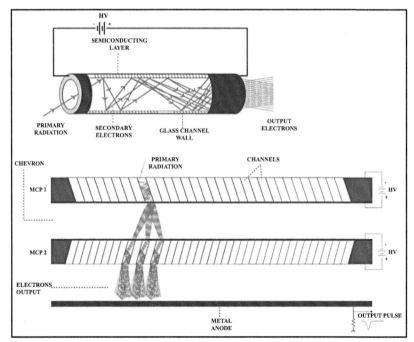

**FIGURE 1.6** Secondary electron multiplier (SEM) and channel electron multipliers (CEM).

MCPs are renowned for their speed and user-friendly manner. However, their basic detection parameters depend solely on the secondary electron emission, which shows discrimination for slower ions. In other words, a decrease in sensitivity is observed with the reduction in the acceleration voltage and an increase in the mass of ions. Irrespective of the technology, almost all the above-mentioned detectors drop sensitivity towards slower ions comparing with the faster ions and subsequently lose the intensity with an increase in mass. In most of cases, this can be identified as the less intense higher mass ions (slower). Post acceleration detectors are introduced against this scenario. Here the ions are accelerated just before the detectors and make sure that the ions hit the first dynode. A more enhanced way of this process is to use conversion dynodes, which are the electrodes kept under high potential to attract the ions from the analyzer, and allows the generation of the first set of secondary ions from their self. Detectors based on secondary ion emission, such as conversion dynodes, can be placed in front of SEM or CEM to obtain good sensitivity at higher molecule masses. They would attract the ions from the mass analyzer and act as an intermediate pusher for both positive and negative ions [16].

Cryogenic detectors came into the picture, giving special emphasis for very large-sized and slow-moving molecules. The mechanism behind cryogenic detectors relies on the phonons production in dielectric absorbers or the creation of quasi-particles in superconductors. Cryogenic detectors provide a wide range of accessible masses, while coupling with HRMS, especially with ToF-MS, and thereby improved the detection limits. They provide good energy resolution, which rather useful in identifying the different charge states and ion fragmentations [17]. Moreover, these special characteristics have been found useful in gathering information other than mass and contribute towards the studies on ion-detector interaction formats, internal energy distribution, charge discrimination, etc., [18].

Focal plane detectors (FPD) find an advantage in simultaneously detecting a range of mass, than focusing a single m/z value.

Another type of detector used in FTMS or in orbitrap is a pair of metal surfaces, placed inside the region of mass analyzers, and allows the ions to pass near them during oscillation. Here, not any direct current is produced, but a weak AC image current is generated between the electrodes.

Different types of detectors allow suitable selection according to the purpose of analysis. In addition, for proper and meaningful results, the selection of proper ionization methods and detectors matters to a great extent.

## 1.7 VACUUM

Mass is always defined separately from the weight, though they are related closely. Their differentiation is associated by the presence of the term 'gravity' in the definition of weight. So, without any doubt, mass can be identified accurately and easily in a system under vacuum, and vacuum furnishes an important part of the MSs.

## 1.8 DATA ACQUISITION

Soft ionization techniques most probably result in the data of intact molecular species (quasi-molecular ions). The quasi-molecular ions alone are not adequate for utilizing in structural characterization. Induced fragmentation by means of collision induced dissociation (CID) or collisionally activated dissociation (CAD) can be conducted to get two different types of spectral data-tandem mass spectrometry and sequential mass spectrometry.

The tandem MS may be the most utilized mode of mass equipment, and its operating principle can be understood by outlining the well-known triple quadrupole (QQQ or TQ). Normally, three kinds of scanning data can be produced by TQ: (i) Product ion scanning, in which, the first quadrupole act as a specific scanner for the precursor ion, and they are transferred to the second quadrupole, which acts as a collision cell, where, the precursor ion is subjected to CID. The fragments produced as a result of CID is scanned by the third quadrupole, resulting in 'product ion spectrum;' (ii) Precursor ion scanning, where the scan is conducted by keeping the third quadrupole static (only one specific product ion is selected) and first quadrupole scanning in a given mass range, to result in a spectrum of precursor ions, which do fragment into same product ion, during CID. The common structural cores/common compound class can be identified by this experiment; (iii) Neutral loss scanning, where both the first and third quadrupoles are kept in scanning mode and the combination of scans provide the data of precursor ions which undergoes CID fragmentation by means of a specific neutral loss. This experiment helps to identify the functional groups.

Mass spectrometry is enormously used for characterization studies, and multistage (sequential) mass spectrometry data furnish even more detailed information in this regard, especially useful for complex samples.

Instruments like IT, FTICR, etc., provides an opportunity for the isolation and re-fragmentation of product ions produced by MS/MS, resulting in MS3 data. The process series of "isolation-re-fragmentation" can be continued in a controlled manner to produce $MS^n$ data where, 'n' represents the number of times, re-fragmentation has been carried out. Structure specific product ions are produced and their interpretation can lead to detailed characterization.

## 1.9 HRMS: SOME SMALL PITFALLS

HRMS encourages ionizable compounds only. Neutral compounds or the poorly ionized ones act transparent and cannot be detected in the system. In the same manner, some of the compounds in the sample need different analyzing conditions than the others (such as ionization modes, pH, voltages, etc.). Therefore, it is necessary to run the sample in the different possible conditions in order to obtain considerable information.

## 1.10 CONCLUSION

Instruments play prime roles in many analytical purposes. While considering the phytochemical analyzes, different instruments are tooled from the very first step of raw material processing up to the final stages of compound identification and structure elucidation. In addition, the proper selection, use, and maintenance of instruments decide the quality of acquiring data.

## KEYWORDS

- atmospheric pressure chemical ionization
- desorption atmospheric pressure photoionization
- fast-atom bombardment
- ionization
- orbitrap
- time-of-flight

## REFERENCES

1. Beynon, J. H., (1960). *Mass Spectrometry and its Application to Organic Chemistry*, Elsevier: Amsterdam.
2. Marshall, A. G., & Hendrickson, C. L., (2008). High-resolution mass spectrometers. *Annu. Rev. Anal. Chem., 1*, 579–599.
3. Fjeldsted, J. C., (2016). Advances in time-of-flight mass spectrometry. In: Perez, S., Eichhorn, P., & Barcelo, D., (eds.), *Comprehensive Analytical Chemistry Applications of Time-of-Flight and Orbitrap Mass Spectrometry in Environmental, Food, Doping, and Forensic Analysis* (Vol. 71, p. 19). Elsevier: Amsterdam.
4. Cameron, A. E., & Eggers, D. F., (1948). An ion "velocitron." *Rev. Sci. Instrum., 19*(9), 605–647.
5. Mamyrin, B. A., (2000). Time-of-flight mass spectrometry (concepts, achievements, and prospects). *Int. J. Mass Spectrom., 206*(3), 251–266.
6. Lee, J., & Reilly, P. T. A., (2011). Limitation of time-of-flight resolution in the ultra-high mass range. *Anal Chem., 83*(15), 5831–5833.
7. Rajski, L., Gómez-Ramos, M. M., & Fernández-Alba, A. R., (2014). Large pesticide multi-residue screening method by liquid chromatography-orbitrap mass spectrometry in full scan mode applied to fruit and vegetables. *J. Chromatog. A., 1360*, 119–127.
8. Comisarow, M. B., & Marshall, A. G., (1974). Fourier transforms ion cyclotron resonance spectroscopy. *Chem. Phys. Lett., 2*(25), 282–283.
9. Kingdon, K. H., (1923). A method for the neutralization of electron space charge by positive ionization at very low gas pressures. *Phys. Rev., 21*(4), 408.
10. Zubarev, R. A., & Makarov, A., (2013). Orbitrap mass spectrometry. *Anal. Chem., 85*, 5288–5296.
11. Zhang, L. K., Rempel, D., Pramanik, B. N., & Gross, M. L., (2005). Accurate mass measurements by Fourier transform mass spectrometry. *Mass Spectrom. Rev., 24*(2), 286–309.
12. Takáts, Z., Wiseman, J. M., Gologan, B., & Cooks, R. G., (2004). Mass spectrometry sampling under ambient conditions with desorption electrospray ionization. *Science, 306*(5695), 471–473.
13. Gross, J. H., (2014). Direct analysis in real time: A critical review on DART-MS. *Anal. Bioanal. Chem., 406*(1), 63–80.
14. Haapala, M., Pól, J., Saarela, V., Arvola, V., Kotiaho, T., Ketola, R. A., Franssila, S., et al., (2007). Desorption atmospheric pressure photoionization. *Anal. Chem., 79*(20), 7867–7872.
15. Kind, T., & Fiehn, O., (2010). Advances in structure elucidation of small molecules using mass spectrometry. *Bioanal. Rev., 2*(1–4), 23–60.
16. Gross, J. H., (2004). Instrumentation. In: *Mass Spectrometry: A Textbook* (pp. 111–192). Springer International Publishing: Berlin.
17. Kraus, H., (2002). Cryogenic detectors and their application to mass spectrometry. *Int. J. Mass Spectrom., 215*(1–3), 45–58.
18. Frank, M., (2000). Mass spectrometry with cryogenic detectors. *Nucl. Instrum. Meth. A., 444, 1*(2), 375–384.

# CHAPTER 2

# Data Processing and Computational Techniques

JOBY JACOB, ANJANA S. NAIR, and SREERAJ GOPI

*Research and Development (R&D) Center, Plant Lipids (P) Ltd.,*
*Kadayiruppu, Kolenchery, Cochin, Ernakulam, Kerala – 682311, India*

## ABSTRACT

There are many processes and procedures are developed for data acquisition. Likewise, newer and newer technologies were developed for their processing also. The properties and methodologies of data processing depend on many factors such as the mode of information, analysis timelines, nature of handling analyte, etc., the chapter considers a discussion on these factors. In addition, different methodologies and processing methods are introduced and explained with relevant examples.

## 2.1 INTRODUCTION

Over the last few decades, phytochemical researches are expanding due to the increase in the incorporation of computational techniques, mathematical modeling, and artificial intelligence. Computational phytochemistry (CP) is an emerging branch of phytochemistry, which efficiently uses computational techniques, mathematical and statistical models to deal with various aspects of phytochemical research. Computer-aided approaches are introduced to save time and money in phytochemical research, such as the identification of metabolomes by the bioactive compound discovery [1]. The impact on phytochemical research by computational methods is visible in recent publications, and will be facilitated in oncoming years since the way we perform phytochemical research today [2]. Since the introduction of combinatorial chemistry and compound libraries favors phytochemical research, both in industry and in academia, it has been diverted to the production of dereplicated phytochemical libraries for HTS for new drug discovery [3].

For conducting dereplication and structural analysis, it is important to understand/process the data acquired by instrument, using an appropriate tool. For hard ionization techniques dealing with high-energy ionization sources, universally accepted databases and libraries are produced, which produce a highly uniform fragmentation pattern of molecules, between instruments. Quasi-molecular ions and their intended fragments are produced from soft ionization techniques; thus, it is not possible to repro-duce the fragmentation pattern or mass signals, between instruments or laboratories. Researches for the reproducible MS spectral data from $MS^2$/ $MS^n$ information have ended up in spectral repositories, by making use of recent technologies such as tuning protocols [4], fragmentation energy index [5], etc. Many attempts have been made to create and validate a standard format for the MS data, so that it could be shared among science communities and be used for data processing. GPMDB [6], PRIDE [7], Peptide Atlas [8], Tranche [9], mzML [10], were some important, successful attempts among them. In spite the success of their accessibility and openness, they were having some drawbacks in terms of feasibility and multiplicity. A Forensic toxicology library was created for LC-QToF containing 56 natural toxic compounds, giving special emphasis on toxicological analysis [11] and it could not contribute to natural products unknown investigations. Different mass spectral analysis tools, which are proved to be potential platform for HRMS data analysis, are discussed below, with selected, relevant examples.

## 2.2   RULE-BASED PREDICTION SYSTEMS

When there is a lack of spectral library, a wise choice is to make use of molecular structural databases. Fragmentation patterns of compounds can be predicted theoretically using the molecular structure database in accordance with fragmentation rules and for the prediction of fragmen-tation there are many commercial tools available. Mass Frontier, ACD/ MS Fragmenter, and MOLGEN-MS are three important systems among them [12]. Researchers have utilized these tools, along with the general fragmentation rules wisely to identify the unknown compounds. Hill et al.

tested 102 compounds, (along with almost 272 other compounds, most of them having the same empirical formula of the test compounds) to determine their chemical structures by matching the experimentally observed CID fragmentation spectra with chemical database fragmentation spectra. Structures of 65 compounds have been identified thus concluding this as a valid method for the structure identification of unknowns [13].

## 2.3 COMBINATORIAL CHEMISTRY

### 2.3.1 COMBINATORIAL LIBRARY

Combinatorial chemistry was originated in the early 1980s. A collection of chemical compounds, small molecules, macromolecules such as proteins which are synthesized by combinatorial chemistry methods are called combinatorial libraries. They are usually represented as one or more structures having a small number of R-group positions, for each there are lists of alternative groups. They were formerly applied to oligonucleotides and peptides which quickly expanded to include synthetic oligomers, small molecules, proteins, and oligosaccharides. Dependent on the type of library desired, they are made and the three fundamental steps of the combinatorial library method are preparation of the library, screening of the library components and determination of the chemical structures of active compounds [3].

Combinatorial libraries can be of two main types, scaffold-based, and backbone-based. In scaffold-based library, the core structure is retained in all compounds in library and variations are done in additional or modified functional groups and in backbone-based library, certain building blocks (e.g., nucleic acids, carbohydrates) are used [3].

### 2.3.2 COMBINATORIAL FRAGMENTATION DATABASE

The combinatorial fragmentation database is developed based on experimental fragmentation spectra and finally matching them to a possible substructure of a known molecular structure, and so, it does not consider the rearrangements at the time of fragmentation [14]. Combinatorial fragmentation starts with assigning cost for bond cleavages, depending on bond energy, bond type and bond dissociation energy. Each peak in the spectra was allocated in the order of minimal cost, to determine the feasibility of bond breakage. Many researchers have conducted to make the enumeration an easy process. A possible and feasible approach was made by constraining the number of cleavages and this concept has inspired the invention of tools such as EPIC [14] and an advanced method named Metfrag [15]. Later, a strategy named MetFusion was introduced by combining Metfrag with a spectral library MassBank. The performance was illustrated using a set of 1062 spectra, achieving a better ranking for the correct compound than while using Metfrag alone [16].

## 2.4 MACHINE LEARNING: PREDICTION OF SUBSTRUCTURE

Instead of depending on the fragmentation patterns, a fabricated prediction framework works well, where the automated classifiers are employed for the prediction and identification of compounds from the spectra. In this method, a set of numerical features are obtained by the classifier transformation corresponding to the molecular properties of the unknown. By comparing it with the feature vectors of the reference compound, it is possible to identify the substructures and compound classes present. Following and modifying the protocol, many approaches were built.

One approach, presented by Heinonen et al. used the predicted feature vectors of LC/MS-CID fragmentation data produced by a kernel-based approach for matching with a huge database such as PubChem. They demonstrated the success of the system by identifying exact structure for almost 65% of the unknowns they have considered [17]. A web server, which can predict the spectra for a chemical structure, annotate peaks in the spectrum of a known chemical structure and identify metabolites from a target spectrum, was prepared by Allen et al. [18].

## 2.5 PREPROCESSING PLATFORM-XCMS

A number of preprocessing platforms are available by now. As a representative of them, let us consider XCMS, a popular, freely available preprocessing approach for LC/MS data of metabolic profiling, which could match and align properties of many samples in a single step. The strategy included: (i) peak detection by applying model peak match filter in sliced LC/MS data; (ii) peak matching, using an algorithm with fixed interval-overlapping bins; and (iii) retention time (RT) alignment by fixing the temporary standards and constructing RT deviation contour. XCMS was demonstrated by analyzing 238 plasma samples (all the samples in duplicate) for metabolic study and 6 spinal cord and brain tissue samples for fatty acid amide hydrolase (FAAH) knockout study. This method is good for huge data handling, discovering new comparisons across samples, and aligning chromatographic traces [19]. Another preprocessing pipeline developed by Falcetta et al. by setting up a mass range gate using the mass difference between the compound and its 5d derivatives [20]. For analyzing LC–MS data software is used to name as "MetSign," where for spectrum deconvolution and peak list alignment a set of data preprocessing algorithms were developed. Due to their ability to provide solutions for peak detection, visualization, peak list alignment, normalization, metabolite putative assignment, and clustering, it can process both the LC-MS and DI-MS data. Another unique feature of MetSign is its ability to analyze the stable isotope labeled data compared to existing bioinformatics tools (Figure 2.1) [21, 22].

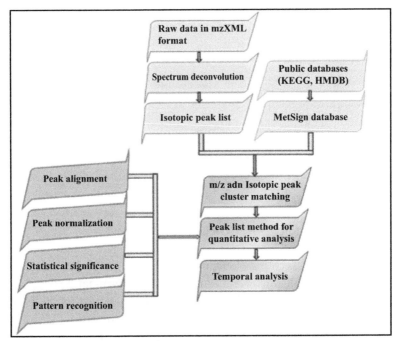

**FIGURE 2.1**  Workflow of "MetSign."

For preprocessing flow injection analysis (FIA)-HRMS raw files and to generate the table of peak intensities "proFIA" software is used which provides innovative algorithms for the purpose. The workflow consists of 3 steps, at first a noise is estimated followed by detection of the peak and finally the peak is quantified, secondly peak grouping was done across samples and finally the missing values imputation is done (Figure 2.2). New indicators which quantify the potential alteration of the feature peak shape due to matrix effect have been implemented by Delabriere et al. In their method preprocessing were fast (less than 15s per file), and main parameter values were easily inferred from the mass resolution of the instrument [23].

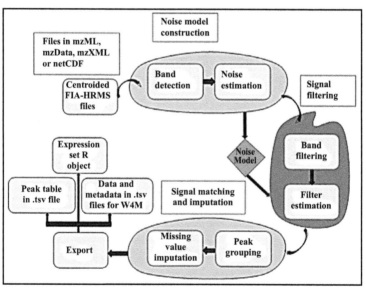

**FIGURE 2.2**   Workflow of "proFIA."

## 2.6  WORKFLOWS AND ALGORITHMS

High-resolution full scan mass spectrometric (HRFSMS) data acquisition utilized the MS-bioinformatics platforms for the determination of the mass and RT of peptides. For synthesizing general data from biospecimens, May et al. introduced an accurate mass and time (AMT) workflow. They have advanced one-step more towards the identification of the proteins and determination of its abundance by utilizing the AMT methodology [24].

A workflow management system, Taverna was introduced and validated, and its query and access on MS proteomics data was explained for a representative repository database-PRIDE [25]. An all ion fragmentation (AIF) approach was considered for the identification of metabolites. A specific in-house mass spectral library was developed by considering accurate mass, MS/MS spectrum, RT, and the product-precursor ion intensity ratios as the control points, which contain 413 compounds [26]. A recent study presented development of an in-house library of flavonoids from the UHPLC-DAD-HRAM-MS$^n$ data. The program proceeds with two steps: (i) identification of the flavonoid classes by UV-Vis spectra; and (ii) identification of individual flavonoids from the mass spectral data. The program was validated and the identification accuracy was found to be 88% [27]. Wang and his team intelligently combined the possibilities of combinatorial fragmentation and algorithms to process the metabolite spectrum match, using their database-searching algorithm named MIDAS [28].

To build, run, and share workflows, Workflow4Metabolomics (W4M; http://workflow4metabolomics.orgonline) infrastructure (W4M e-infrastructure) offers a user-friendly and computationally efficient environment. The W4M 3.0 release provides a total of 40 tools, from preprocessing, statistical analysis, and up to annotation. Advanced LC-HRMS analysis, GC-MS, FIA-MS, and NMR workflows can be easily done since it comes up with new modules [29]. MZmine, an open-source software toolbox for LC-MS data processing was first introduced in 2005. It implemented a simple method for data processing and visualization. The software has been applied to numerous metabolomic analyzes and is also compared with related software packages. MZmine 2 can be downloaded from the project WWW site along with its accessories. It is being applied in metabolomic researches since the current version can process large batches of data for targeted and nontargeted analyzes [30].

## 2.7 RETENTION TIME (RT) PREDICTION

A comparatively feasible aid for compound identification is to exploit the RT. A database of RTs can be shared across the chromatographic systems easily. An RT prediction model was developed by Creek et al. by applying a multiple linear regression with six physicochemical parameters of 120

standard metabolites. It could eliminate more than 40% of false identifications [31]. Combination of two approaches named retention projection
and back-calculation resulted in a more accurate method for calculating
LC retention between gradients and labs [32]. One proposed model of RT
database includes the RT's from different chromatographic systems and
it provides a projection model between the RT's of all possible pairs of
chromatographic systems in the database [33].

The relation between RT and the chemical structure is studied by
Quantitative structure-retention relationship (QSRR) tool. A linear solvent
strength (LSS) theory was used to prepare QSRR models in order to
predict the RT, under any gradient conditions [34]. The same group has
come forward with an advanced dynamic database for human metabolites,
by employing the same platform [35]. Another proposed model for RT
prediction, based on the chemical structures has designed from hydrophobic subtraction model (HSM) and QSRR [36].

## 2.8   MASS SPECTRAL TREES

Fragmentation trees are created from the acquired spectral hierarchy data
of sequential fragmentation ($MS^n$). In one study, fragmentation reactions efficiently represented the fragmentation, where the nodes stand
for fragments and the directed edges represent the reactions between
them [37]. The created 'cheminformatical' tool-multistage elemental
formula generator (MEF)—was checked for its capability of chemical
similarity searching, by using a fingerprint-based algorithm named
Tanimoto coefficient [38]. Two different $MS^n$ libraries were created
and their performance was validated by the novel semi-automatic de
novo identification tool. They have provided a strong platform for the
identification for both known and unknown compounds, along with
the structure elucidation of unknowns. Later, the same research group
came forward with a follow up for the spectral tree-a pipeline, which
utilizes the $MS^n$ data of unknown metabolites for the identification of
the same [39].

Another algorithm named precursor ion fingerprinting (PIF) was
developed for the interpretation and library search of the mass spectra,
were structurally related alkaloids were used to demonstrate them [40].
A spectral tree was constructed from the $MS^n$ data of these alkaloids.

The ion substructures, identified for the spectral data is processed with the spectral library to obtain the corresponding potential structure. Thus, a universal $MS^n$ spectra library was created to utilize for the identification and structure elucidation of unknown compounds. In one study, the $MS^n$ data is annotated with hierarchical trees of the analyte and the approach was demonstrated with the dataset of green tea extract [36].

Altogether, HRMS techniques provide a much better opportunity to get data of the sample. They can provide a complete overview of the product by allowing compound identification, profiling, fingerprinting, dereplication, metabolite phenotyping, etc. The complete structural characterization can help even the structure-based drug design and development. The current review gives special emphasis on the use of High-resolution mass spectrometry compatible with LC in the field of identification, characterization, and quantification of bioactive constituents from phytochemical extracts and products. Some essential practical examples are presented to demonstrate the same.

## 2.9 METABOLOMIC TOOLS

Metabolomics is the large-scale study of metabolites and metabolome is a term used for representing metabolites and their interactions within a biological system. It is in the early 70s, metabolite-profiling publications originated from the Baylor College of Medicine [41–43]. At that time, GC/MS was used by these authors to illustrate their concept through the multicomponent analyzes of steroids, and neutral and acidic urinary drug metabolites. They are also coined the term "metabolite profiling" to refer qualitative and quantitative analyzes of complex mixtures of physiological origin [44]. The steps of analysis of metabolomics data or computational metabolomics can be divided into three steps, raw data is preprocessed to the sample by a variable matrix of intensities, followed by the statistical analysis which detects variables of interest and build prediction models, and finally annotation of variables to provide insight into their chemical and biological functions. We can perform statistics and annotation steps in the reverse order to get a first-pass overview of the dataset content by performing an automatic query of metabolite databases [29].

### 2.9.1 IMAGING MASS SPECTROMETRY (MSI) TOOLS

Resurgence in matrix-assisted laser desorption/ionization (MALDI)-MS-based, desorption electrospray ionization (DESI), laser desorption/ionization (LDI), and laser ablation electrospray ionization (LAESI)-MS-based tissue imaging has led to new software for computing and processing data. "Spectral Analysis" is a software which can be used through the entire analysis workflow, since it provides a wide range of methods for normalization, smoothing, baseline correction, and image generation to multivariate analysis such as non-negative matrix factorization (NMF), memory efficient principal component analysis (PCA), maximum autocorrelation factor (MAF), and probabilistic latent semantic analysis (PLSA), for data sets acquired from single experiments to large multi-instrument, multimodality, and multicenter studies [45].

### 2.9.2 NUCLEAR MAGNETIC RESONANCE (NMR) BASED TOOLS

The importance of multidimensional NMR and hybrid MS/NMR methods in deciphering known and unknowns in metabolomics is discussed worldwide. It is important to have valuable software for processing and visualizing data in the research field. Commonly used software includes Campus Chemical Instrument Center (CCIC), NMRPro, SpiNCouple, Chemical Shifts to Metabolic Pathways (ChemSMP), PROMED, SpeckTackle, NMRmix, and jsNMR. "CCIC" has evolved as an immensely useful resource of NMR-based metabolomics data analysis and metabolite identification tools (http://spin.ccic.ohio-state.edu/) which contains multiple tools such as Covariance, 1D NMR Query, 2D 13C-1H HSQC Query, 2D 13C-TOCCATA Query, 2D 1H(13C)-TOCCATA Query, DemixC, COLMAR, and Multiple spectra Query [25].

"NMRPro," a user-friendly web component can be easily incorporated into currently used web applications by enabling a Python package, managed by Django App and SpecdrawJS based online interactive process and visualization [46]. "SpiNCouple" is software having a two-dimensional (2D) 1H-1H J-resolved NMR database from 598 metabolite standards as the backbone [47]. It provides the spectra that include both J-coupling and 1H chemical shift information allowing spectral annotation, especially for metabolic mixtures and options for absolute-quantitative analysis. Without individual metabolite identification, "ChemSMP" approach identifies active metabolic pathways directly from chemical shifts obtained from a single 2D [13C-1H] correlation NMR spectrum to facilitate rapid pathway mapping analysis [48].

"PROMED" an NMR-based metabolomics tool works in the basis of matching the data of the pattern of peaks rather than absolute tolerance thresholds by using a combination of geometric hashing and similarity scoring, thus helps in compound identification and assignment of metabolites independent of the pH, temperature, and ionic strength techniques [49]. Cross-browser compatible software for spectra visualization "Speck-Tackle," is a custom-tailored JavaScript charting library for spectroscopy data in life sciences which contain library of several default chart types and supports common functionality like spectra overlays or tooltips.

"NMRmix" is a tool that supports the creation of ideal mixtures to optimize the composition of the mixtures to minimize spectral peak overlaps from a large panel of compounds with known chemical shifts using a simulated annealing algorithm. A graphical user interface simplifies data import and visualization [50]. "jsNMR" is a lightweight NMR spectrum viewer providing a cross-platform spectrum visualizer which runs on all computer architectures (including mobile devices) [51]. jsNMR allows for conversion of spectrum data to alternative file formats such as SIMPSON spectrum.csv file and PDF format, in addition to visualization.

### 2.9.3 TARGETED AND UNTARGETED ANALYSIS TOOLS

Targeted analysis tools are milestones for analyzing specific metabolic perturbations under systems of interest by using high-resolution instruments. For visualization, data interpretation, and processing, it is important to have efficient software since large-scale targeted metabolomic studies remains an area of considerable interest. Commonly used software available for targeted metabolomic studies were "RIPPER," "MRMAnalyzer," and "MetDIA." For MS-based label-free relative quantification for metabolomics and proteomics studies "RIPPER" is used. It has features such as data pre-processing, RT alignment, analyte quantification, and grouping across runs [52]. For automatic rapid processing of large set of multiple reactions monitoring (MRM) based targeted metabolomics data, "MRMAnalyzer," an R package were used [53]. Data processing steps include peak detection and alignment, 'pseduo' accurate m/z transformation, check for quality control (QC), identification of metabolite and statistical analyzes.

For efficient data-independent acquisition (DIA) data analysis "MetDIA" approach was implemented which allowed targeted extraction of metabolites from multiplexed MS/MS by considering each metabolite in the spectral library as an analysis target. MetDIA allows for detection of ion chromatograms for each metabolite (both precursor ions and fragment ions) in $MS^2$ data along with detection, extraction, and scoring metabolite identifications which is referred as metabolite-centric identification.

Untargeted metabolomics has gained considerable popularity allowing for expanded coverage of metabolites in matrices of interest. However, it shows some difficulty in processing data and is tried to solve by using online tools such as Intelligent Metabolomic Quantitation (iMet-Q), nontargeted diagnostic ion network analysis (NINA), MSCombine, and msPurity [54].

"iMet-Q" is a software tool like RIPPER, which performs peak detection and peak alignment, which provide a summary of qualitative results along with reports on ion abundance at both sample and replicate levels. The software also provides detected metabolite peak charge states and isotope ratios for facilitating metabolite identification [55]. Another software NINA shows its ability to summarize all of the fragment ions from the acquired MS/MS spectra which were shared by the precursors and performs post-data acquisition analysis there by determining the non-targeted diagnostic ions (NIs). Once a single compound has been identified de *novo* NI-guided network using bridging components with two or more NI can be established which can be utilized for sequential identification of the structures of all NIs [56].

## 2.9.4 ANNOTATION

Metabolite identification or annotation remains a major focal point and area of extensive investigation in untargeted metabolomics research. Recently, several new approaches have been developed to facilitate the identification of unknowns, which are discussed here. Spalding et al. introduced an alternative approach for the identification of unknowns called barcoding MS that is not reliant on full, high-resolution MS/MS spectra [57]. In addition to barcoding of MS$^2$ spectra, for facilitating identification of unknown, string-based regular expressions of MS/MS spectra can be considered [58]. "iMet" is a web-based computational tool based on experimental tandem mass spectrometry that allows annotation of unknown metabolites. "iMet" MS/MS spectra that identify metabolites structurally similar to an unknown metabolite through a net atomic addition or removal that converts the known metabolite into the unknown one [59].

Plant metabolite annotation toolbox (PlantMAT) is an Excel-based software used for the prediction of plant natural products such as glycosylated flavonoids and saponins through combinatorial enumeration of aglycone, glycosyl, and acyl subunits using informed phytochemical knowledge. For metabolite annotation, the custom software allowed operation of an automated and streamlined workflow which has a user-friendly interface within Microsoft Excel. It also increased the chemical and the metabolic space of traditional chemical databases [2, 44]. "FlavonQ," an automated data processing approach/workflow is specifically oriented towards profiling of flavone and flavonol glycosides with ultra-high-performance liquid chromatography-diode array detection-high resolution accurate mass-mass spectrometry (UHPLC-DAD-HRAM-MS). It performs data format conversion, peak detection, flavone, and flavonol glycoside peak extraction and identification, and generation of quantitative results [60]. For the analysis of glucosinolates (GSL) using UHPLCHRAM/MS$^n$ technology, the MATLAB-based "GLS-Finder," was developed. GLS-Finder is capable to facilitate both qualitative and semiquantitative analyzes of GSL through [raw data deconvolution, peak alignment, glucosinolate putative assignments, quantitation, and unsupervised PCA [44].

## 2.10 PLATFORM FOR PLANT METABOLOMICS: AUTOMATIC DATA ANALYSIS

### 2.10.1 A FIVE MODULE APPROACH FOR UHPLC-HRMS

Analytical hardware's are capable of capturing robust data sets for many types of biological samples. However, it is a challenge to accurately extract qualitative and quantitative information of number of metabolites using UHPLC-HRMS thereby affecting metabolomic and lipidomic field [61].

For overcoming these drawback, there are a number of proprietary and freely available methodologies are developed, including AntDAS [62], Mzmine2 [63], XCMS [64] and MS-DIAL [65]. For untargeted metabolomics such as EIC construction, detection of peak, annotation of peak and peak alignment, these methodologies typically integrate the entire data analysis workflow and connect to chemometrics methods. These chemometric methods screen out functionally impactful metabolites which exhibit significant differences amongst various experimental groups.

Researchers are still faced with many challenges in practical applications of these methodologies. There are possibilities to show same biological metabolite for identifying signals, especially for complex plant sample due to the screen out of hundreds of ions based on analysis of variance (ANOVA) or partial least square chemometrics. As a solution to this problem researcher have to identify ions that putatively originate from a single metabolite especially neutral loss and fragment ions manually and off course it is certainly a very time-consuming task. To address this problem Liu et al. developed an ion clustering-based fragment identification algorithm with their previously developed data analysis methods [66, 67]. This includes peak detection, time shift, correction and registration modules which provide an integrated data analysis platform for UHPLC-HRMS based untargeted metabolomics. The platform comprises of five modules:

- EIC peak extraction;
- time shift correction;
- peak registration across samples;
- peak screening module; and
- ion clustering-based peak annotation module [68].

### 2.10.1.1   EIC PEAK EXTRACTION MODULE

In this module, the first step is to transform acquired UHPLC-HRMS data files from an instrument into the mzXML file format by using "ProteoWizard" software. By using an ion density clustering algorithm [67] and considering the fact that the ions from a metabolite within a small m/z tolerance (0.01 Da) exhibit almost identical m/z values, EICs are constructed where the specific ion density will be higher than any specific background noise signal thereby performing the peak detection for each EIC.

At first, for baseline correction, a local-minimal value-based baseline drift correction algorithm is introduced. It is followed by the extraction of chromatographic peaks using a Gaussian smoothing-based strategy [66]. Peaks are extracted by smoothing the EIC under different smoothing scales. The next step is to search the ridgelines across successively increased smoothing scales. To characterize each EIC, peak RT and m/z value of the ion acquired at the peak apex are used. Finally, the quantitative information on peak height and area is extracted by using the sum of responses in the elution EIC peak and intensity at the peak apex [68].

### 2.10.1.2  TIME SHIFT CORRECTION MODULE

It is important to perform time shift correction for each EIC. For selecting the reference sample, a total number of peaks in each EIC is counted, and EIC with a maximum number of components is chosen, and it is important that each peak must satisfy m/z and retention tolerances (0.01 Da and 0.5 min respectively). A similarity matrix is constructed by checking the similarity between a test peak and a reference peak using Pearson correlation coefficient values of EIC curves. Based on maximizing accumulated Pearson coefficients in the matrix, Liu et al. aligned all reference peaks simultaneously by searching an optimization path using a modified dynamic programming (DP) [62]. Aligned peaks are represented as Nodes, on its basis the time shift values for a given EIC are estimated using the linear interpolation function in MATLAB thus resolving time shift problem accurately [68].

### 2.10.1.3  PEAK REGISTRATION MODULE AND PEAK SCREENING MODULE

Peak registration module classifies EIC peaks corresponding to the same ion from different samples into a group of single ion. Liu et al. developed a nearest neighboring connecting (NNC) algorithm [67] to register EIC peaks that uses the modified RT of each EIC peak obtained from the previous module. According to their RT differences, candidate peaks are presented in an ascending order and each connected pair is scanned by the NNC algorithm. A group cannot include more than one EIC peak from the same sample. Until EIC peaks cannot be classified further, the scanning procedure of NNC iteratively repeats. Finally, a registered component table is made where EIC peaks in an NNC group are registered with the same identifier.

In this module, metabolites are first screened and peaks are identified with statistically significant differences in mean values amongst sample groups using ANOVA. Using obtained bilinear structure ions from a single metabolite can be identified by linear correlation of peak height or peak area values. Peak heights with a minimum covariance determined by Pearson coefficient between two screened peaks is used to eliminate the influence of outlying samples. If the coefficient between two screened ions is above a user-defined value (for ex. 0.9), ions will be temporarily identified and are marked with an identical "Meta ID" [68].

## 2.10.1.4   ION CLUSTERING-BASED PEAK ANNOTATION MODULE

In this module, it is verified that the ions with identical 'Meta ID' come from the same metabolite or not. Liu et al. developed a bottom-up ion-clustering algorithm that iteratively identifies the ions that are supposed to originate from a single metabolite. Its first step is to identify isotopic ions (for example $[E+H]^+$ and $[E + 1 + H]^+$) from a putative compound on the basis of an initialized EIC peak shape similarity cut-off value for example 0.9 and RT tolerance for example 0.02 min.

EIC peak shape similarity cut-off value for each $[E+H]^+$ ion will be adaptively determined on the basis of its isotopic ions (such as $[E + 2 + H]^+$, $[E + 3 + H]^+$) and/or adduct ions ($K^+$, $NH_4^+$, $Na^+$, etc.), are recognized. $[E+H]^+$ ions are clustered to identify its origin using their peak shapes and RT. To verify the identical nature or identical 'Meta ID' of ions, the platform returns to the peak screening module. Clustered ions from sample, samples are identified and a derived mass spectrum for the metabolite is finally generated [62, 63, 68].

By using these five modules, it is easy to accurately identify fragment ions from same metabolite through automated UHPLC-HRMS data analysis. However, while comparing with the publicly available untargeted metabolomics data analysis tools, some differences are noticed for this tool that, it doesn't require data transformation amongst the different modules for fragment ion identification. MATLAB GUI is freely available software used for fragment ion identification [68].

## *2.10.2   XCMS ONLINE*

A number of preprocessing platforms are available by now. Unlike the web-based tools recently introduced to perform statistical analysis of preprocessed data (MetaboAnalyst and metaP-Server), XCMS Online serves as a solution for the entire untargeted metabolomics workflow [69]. The strategy of XCMS included: (i) peak detection by cutting the LC/MS data into slices and applying model peak match filter; (ii) peak matching across the samples, using an algorithm with fixed interval-overlapping bins; and (iii) RT alignment by fixing the temporary standards and constructing RT deviation contour. The information such as peak areas of all detected metabolite features across analyzed samples and pooled QC sample are arranged as a data table in.txt format.

The metabolites are assigned and annotated with CAMERA for isotopes using the couple of specific RT and accurate m/z ratio information. The software detects the peak and further integrates there by correcting metabolites abundances for signal drift effect by fitting a locally quadratic regression model to the QC values. For offline analysis and publication, results and images are downloaded as zip files [70].

### *2.10.2.1   WORK FLOW OF XCMS*

A three-step organization of metabolomic data processing is the primary step in XCMS online software, which includes data upload, parameter selection, and result interpretation (Figure 2.3). User can simply drag and drop their files in the accepted file format through a specific Java applet for sample comparison. The accepted file formats are netCDF, mzData, mzXML, Agilent.d folders [69]. For example, in some analysis the raw LC-MS data were first converted to mzXML files using a convert like ProteoWizard MS Convert and then uploaded to XCMS software [70]. All files are automatically compressed and encrypted through a secure SSL connection before being uploaded. The job can be submitted even before the upload is complete since the file upload will continue in the background and user doesn't need to wait for the upload to finish. The data processing will start automatically after the successfully uploaded of all files.

During the upload of a file, users will be asked to select a parameter set which matches the instrument setup and each instrument has a predefined parameter sets for different instrument setups like HPLC/Q-ToF, UHPLC/Q-ToF, HPLC/Orbitrap, HPLC/single quad MS, GC/single quad MS. Customization of parameter sets can be done in order to change the signal or noise threshold, to adjust the mass tolerance of the identification step, or feature detection methods are to be changed, RT correction, alignment, and annotation. The job will be submitted to the system after parameter set selection. Data processing can take from minutes up to hours depending on the data set size while multiple jobs can be submitted simultaneously. Proper management of all data sets that have been previously uploaded is an important feature of this system, which can be utilized for additional comparisons, modified, or deleted [69].

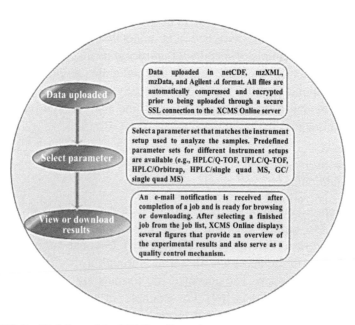

**FIGURE 2.3**   Workflow of the XCMS online software.

## 2.10.3 RESULT INTERPRETATION

After a job has been completed, the user receives an email notification, which shows it is ready for browsing or downloading results. Several figures providing an overview of the experimental results are displayed in XCMS online, which serve as a QC mechanism. Non-linear methods are used by XCMS Online for compensation of RT drifts between samples. Before and after the correction of RT, an overlay of all total ion chromatograms (TICs) acquired is shown as the visualization and QC of this correction procedure, in addition to the RT correction curves. All TICs are aligned after RT correction and recognized potentially problematic samples with extreme deviations can be removed from the data set.

Dysregulated features representing ions whose intensities are altered between sample groups are plotted as "mirror plot" which shows data according to statistical thresholds set by the user for example, p-value ≤0.001, and fold change ≥2. Features that are upregulated are represented as circles in green on the top and features that are down regulated are represented as circles in red on the bottom. The size of each circle corresponds to the fold (log) change of the feature which is having an average difference in relative intensity of the peak between sample groups. Greater fold changes are represented as larger circles and p-values are represented by shades of color where brighter circles show lower p-values. The TICs whose RT is corrected are overlaid in gray in the background of the figure and with a black outline; the circles representing features with hits in the METLIN database are shown. An interactive version of the plot is available when users scroll their mouse over the circles in the plot featuring statistics and putative identities displayed in a pop-up window.

For visualization of high-dimensional data sets, two additional plots are also included namely a multidimensional scaling (MDS) plot and a PCA plot which are performed on the centered and scaled data. The aligned feature table and all graphics along with complete results can be downloaded as a zip file from the overview page. "Browse Result Table" button is used to view the feature table which contain detailed information of individual feature such as statistics, ion chromatogram extracted, details of the spectrum and METLIN assignments.

Multiple criteria are available to filter the results, to prioritize features and to facilitate the interpretation. On the basis of fold change, p-value, m/z value, intensity, and RT ranges filtering are done. Isotopic peaks can be removed conveniently. The table, which is filtered and annotated, can be saved in TSV format for import into Microsoft Excel or other programs [69].

## 2.11  CONCLUSION

Analyzes of phytochemicals are done by using a set of data preprocessing techniques by different hyphenated HRMS instruments. It includes rule-based prediction systems, machine learning-prediction systems data preprocessing techniques, metabolomics tools, etc. Online methodologies are ease to use and user-friendly. In this chapter, we have discussed different data processing methods and its relation to metabolomics, functional genomics, and systems biology. Many of the technologies such as optical spectroscopy, nuclear magnetic resonance (NMR), and mass spectrometry are mentioned briefly. The important role of bioinformatics and various data visualization methods are assessed and summarized.

## KEYWORDS

- **combinatorial libraries**
- **machine learning**
- **mass spectral trees**
- **substructures**
- **total ion chromatograms**
- **workflows**

## REFERENCES

1. Sarker, S. D., & Nahar, L., (2017). Computer-aided phytochemical research. *Trends Phytochem. Res., 1*(1), 1–2.
2. Sarker, S. D., & Nahar, L., (2018). An introduction to computational phytochemistry. In: Sarker, S., & Nahar, L., (ed.), *Computational Phytochemistry* (p. 1). Elsevier: USA.

3. Nahar, L., & Sarker, S. D., (2018). Application of computation in building dereplicated phytochemical libraries. In: Sarker, S., & Nahar, L., (ed.), *Computational Phytochemistry* (p. 141). Elsevier: USA.
4. Champarnaud, E., & Hopley, C., (2011). Evaluation of the comparability of spectra generated using a turning point protocol on twelve electrospray ionization tandem-in-space mass spectrometers. *Rapid Commun. Mass Spectrom., 25,* 1001–1007.
5. Palit, M., & Mallard, G., (2009). Fragmentation energy index for univerzalization of fragmentation energy in ion trap mass spectrometers for the analysis of chemical weapon convention related chemicals by atmospheric pressure ionization-tandem mass spectrometry analysis. *Anal Chem., 81*(7), 2477–2485.
6. Craig, R., Cortens, J. P., & Beavis, R. C., (2004). Open-source system for analyzing, validating, and storing protein identification data. *J. Proteome. Res., 3*(6), 1234–1242.
7. Martens, L., Hermjakob, H., Jones, P., Adamski, M., Taylor, C., States, D., Gevaert, K., et al., (2005). PRIDE: The proteomics identifications database. *Proteomics, 5*(13), 3537–3545.
8. Desiere, F., Deutsch, E. W., King, N. L., Nesvizhskii, A. I., Mallick, P., Eng, J., Chen, S., et al., (2006). The peptide atlas project. *Nucleic Acids Res., 1*(34), 655–658.
9. Falkner, J. A., & Andrews, P. C., (2007). P6-T tranche: Secure decentralized data storage for the proteomics community. *J. Biomol. Tech., 18*(1), 3.
10. Martens, L., Chambers, M., Sturm, M., Kessner, D., Levander, F., Shofstahl, J., Tang, W. H., et al., (2011). mzML: A community standard for mass spectrometry data. *Mol. Cell Proteomics, 10*(1) 1–7.
11. Ogawa, T., Zaitsu, K., Kokaji, T., Suga, K., Kondo, F., Iwai, M., Suzuki, T., et al., (2020). Development and application of a forensic toxicological library for identification of 56 natural toxic substances by liquid chromatography-quadrupole time-of-flight mass spectrometry. *Forensic Toxicol., 38*(1), 232–242.
12. Hufsky, F., Scheubert, K., & Böcker, S., (2014). Computational mass spectrometry for small-molecule fragmentation. *Trends Anal. Chem., 53,* 41–48.
13. Hill, D. W., Kertesz, T. M., Fontaine, D., Friedman, R., & Grant, D. F., (2008). Mass spectral metabonomics beyond elemental formula: Chemical database querying by matching experimental with computational fragmentation spectra. *Anal. Chem., 80*(14), 5574–5582.
14. Hill, A. W., & Mortishire-Smith, R. J., (2005). Automated assignment of high-resolution collisionally activated dissociation mass spectra using a systematic bond disconnection approach. *Rapid Commun. Mass Spectrom., 19*(21), 3111–3118.
15. Wolf, S., Schmidt, S., Müller-Hannemann, M., & Neumann, S., (2010). *In silico* fragmentation for computer assisted identification of metabolite mass spectra. *BMC Bioinform., 11*(1), 148.
16. Gerlich, M., & Neumann, S., (2013). MetFusion: Integration of compound identification strategies. *J. Mass Spectrom., 48*(3), 291–298.
17. Heinonen, M., Shen, H., Zamboni, N., & Rousu, J., (2012). Metabolite identification and molecular fingerprint prediction through machine learning. *Bioinformatics, 28*(18), 2333–2341.
18. Allen, F., Pon, A., Wilson, M., Greiner, R., & Wishart, D., (2014). CFM-ID: A web server for annotation, spectrum prediction, and metabolite identification from tandem mass spectra. *Nucleic Acids Res., 42*(W1), W94–W99.

19. Smith, C. A., Want, E. J., O'Maille, G., Abagyan, R., & Siuzdak, G., (2006). XCMS: Processing mass spectrometry data for metabolite profiling using nonlinear peak alignment, matching, and identification. *Anal Chem., 78*(3) 779–787.
20. Falcetta, F., Morosi, L., Ubezio, P., Giordano, S., Decio, A., Giavazzi, R., Frapolli, R., Prasad, M., et al., (2018). Past-in-the-future. Peak detection improves targeted mass spectrometry imaging. *Anal. Chim. Acta, 1042,* 1–10.
21. Wei, X., Shi, X., Kim, S., Zhang, L., Patrick, J. S., Binkley, J., McClain, C., & Zhang, X., (2012). Data preprocessing method for liquid chromatography-mass spectrometry-based metabolomics. *Anal. Chem., 84*(18), 7963–7971.
22. Wei, X., Sun, W., Shi, X., Koo, I., Wang, B., Zhang, J., Yin, X., et al., (2011). MetSign: A computational platform for high-resolution mass spectrometry-based metabolomics. *Anal. Chem., 83*(20), 7668–7675.
23. Delabrière, A., Hohenester, U. M., Colsch, B., Junot, C., Fenaille, F., & Thévenot, E. A., (2017). ProFIA: A data preprocessing workflow for flow injection analysis coupled to high-resolution mass spectrometry. *Bioinformatics, 33*(23), 3767–3775.
24. May, D., Fitzgibbon, M., Liu, Y., Holzman, T., Eng, J., Kemp, C. J., Whiteaker, J., Paulovich, A., & McIntosh, M., (2007). A platform for accurate mass and time analyses of mass spectrometry data. *J. Proteome Res., 6*(7), 2685–2694.
25. Jiménez, R. C., & Vizcaíno, J. A., (2013). Proteomics data exchange and storage: The need for common standards and public repositories. In: Matthiesen, R., (ed.), *Mass Spectrometry Data Analysis in Proteomics* (Vol. 1007, p. 317).Humana Press, Totowa, NJ.
26. Naz, S., Gallart-Ayala, H., Reinke, S., Mathon, C., Blankley, R., Chaleckis, R., & Wheelock, C. E., (2017). Development of an LC-HRMS metabolomics method with high specificity for metabolite identification using all ion fragmentation (AIF) acquisition. *Anal. Chem., 89*, 7933–7942.
27. Zhang, M., Sun, J., & Chen, P., (2017). Development of a comprehensive flavonoid analysis computational tool for ultra-high-performance liquid chromatography-diode array detection-high resolution accurate mass-mass spectrometry data. *Anal. Chem., 89*(14), 7388–7397.
28. Wang, Y., Kora, G., Bowen, B. P., & Pan, C., (2014). MIDAS: A database-searching algorithm for metabolite identification in metabolomics. *Anal. Chem., 86*(19), 9496–9503.
29. Guitton, Y., Tremblay-Franco, M., Le Corguillé, G., Martin, J. F., Pétéra, M., Roger-Mele, P., Delabrière, A., et al., (2017). Create, run, share, publish, and reference your LC-MS, FIA-MS, GC-MS, and NMR data analysis workflows with the Workflow4Metabolomics 3.0 Galaxy online infrastructure for metabolomics. *Int. J. Biochem. Cell Biol., 93*, 89–101.
30. Pluskal, T., Castillo, S., Villar-Briones, A., & Orešič, M., (2010). MZmine 2: Modular framework for processing, visualizing, and analyzing mass spectrometry-based molecular profile data. *BMC Bioinform., 11*(1), 395.
31. Creek, D. J., Jankevics, A., Breitling, R., Watson, D. G., Barrett, M. P., & Burgess, K. E. V., (2011). Toward global metabolomics analysis with hydrophilic interaction liquid chromatography-mass spectrometry: Improved metabolite identification by retention time prediction. *Anal. Chem., 83*(22), 8703–8710.

32. Abate-Pella, D., Freund, D. M., Ma, Y., Simón-Manso, Y., Hollender, J., Broeckling, C. D., et al., (2015). Retention projection enables accurate calculation of liquid chromatographic retention times across labs and methods. *J. Chromatogr. A, 18*(1412), 43–51.

33. Stanstrup, J., Neumann, S., & Vrhovšek, U., (2015). PredRet: Prediction of retention time by direct mapping between multiple chromatographic systems. *Anal. Chem., 87*(18), 9421–9428.

34. Randazzo, G. M., Tonoli, D., Hambye, S., Guillarme, D., Jeanneret, F., Nurisso, A., Goracci, L., et al., (2016). Prediction of retention time in reversed-phase liquid chromatography as a tool for steroid identification. *Anal. Chim. Acta., 15*(916), 8–16.

35. Randazzo, G. M., Tonoli, D., Strajhar, P., Xenarios, I., Odermatt, A., Boccard, J., & Rudaz, S., (2017). Enhanced metabolite annotation via dynamic retention time prediction: Steroidogenesis alterations as a case study. *J. Chromatogr. B. Analyt. Technol. Biomed. Life Sci., 15*(1071), 11–18.

36. Ridder, L., Van, D. H. J. J., Verhoeven, S., De Vos, R. C., Bino, R. J., & Vervoort, J., (2013). Automatic chemical structure annotation of an LC-MS (n) based metabolic profile from green tea. *Anal Chem., 18*(85), 6033–6040.

37. Rojas-Cherto, M., Peironcely, J. E., Kasper, P. T., Van, D. H. J. J. J., De Vos, R. C. H., Vreeken, R., Hankemeier, T., & Reijmers, T., (2012). Metabolite identification using automated comparison of high-resolution multistage mass spectral trees. *Anal. Chem., 84*(13), 5524–5534.

38. Kasper, P. T., Rojas-Chertó, M., Mistrik, R., Reijmers, T., Hankemeier, T., & Vreeken, R. J., (2012). Fragmentation trees for the structural characterization of metabolites. *Rapid Commun. Mass Spectrom., 26*(19), 2275–2286.

39. Peironcely, J. E., Rojas-Chertó, M., Tas, A., Vreeken, R., Reijmers, T., Coulier, L., & Hankemeier, T., (2013). Automated pipeline for de novo metabolite identification using mass-spectrometry-based metabolomics. *Anal. Chem., 85*(7), 3576–3583.

40. Sheldon, M. T., Mistrik, R., & Croley, T. R., (2009). Determination of ion structures in structurally related compounds using precursor ion fingerprinting. *J. Am. Soc. Mass Spectrom., 20*(3), 370–376.

41. Devaux, P. G., Horning, M. G., & Horning, E. C., (1971). Benyzloxime derivative of steroids: A new metabolic profile procedure for human urinary steroids. *Anal. Lett., 4*(3), 151.

42. Horning, E. C., & Horning, M. G., (1971). Human metabolic profiles obtained by GC and GC/MS. *J. Chromatogr. Sci., 9*(3), 129–140.

43. Horning, E. C., & Horning, M. G., (1971). Metabolic profiles: Gas-phase methods for analysis of metabolites. *Clin. Chem., 17*(8), 802–809.

44. Sumner, L. W., Mendes, P., & Dixon, R. A., (2003). Plant metabolomics: Large-scale phytochemistry in the functional genomics era. *Phytochemistry, 62*(6), 817–836.

45. Misra, B. B., Fahrmann, J. F., & Grapov, D., (2017). Review of emerging metabolomic tools and resources: 2015–2016. *Electrophoresis, 38*(18), 2257–2274.

46. Mohamed, A., Nguyen, C. H., & Mamitsuka, H., (2016). Current status and prospects of computational resources for natural product dereplication: A review. *Bioinformatic., 17*(2), 309–312.

47. Kikuchi, J., Tsuboi, Y., Komatsu, K., Gomi, M., Chikayama, E., & Date, Y., (2015). SpinCouple: Development of a web tool for analyzing metabolite mixtures via two-dimensional J-resolved NMR database *Anal. Chem., 88*, 659–665.
48. Dubey, A., Rangarajan, A., Pal, D., & Atreya, H. S., (2015). Chemical shifts to metabolic pathways: Identifying metabolic pathways directly from a single 2D NMR spectrum. *Anal. Chem., 87*(24), 12197–12205.
49. Dubey, A., Rangarajan, A., Pal, D., & Atreya, H. S., (2015). A pattern recognition-based approach for identifying metabolites in NMR based metabolomics. *Anal. Chem., 87*(14), 7148–7155.
50. Beynon, J. H., (1960). *Mass Spectrometry and its Applications to Organic Chemistry*. Elsevier, Amsterdam.
51. Vosegaard, T., (2015). jsNMR: An embedded platform-independent NMR spectrum viewer. *Magn. Reson. Chem., 53*(4), 285–290.
52. Van, R. S. K., Higgins, L., Carlis, J. V., & Griffin, T. J., (2016). RIPPER: A framework for MS1 only metabolomics and proteomics label-free relative quantification. *Bioinformatics, 32*(13), 2035–2037.
53. Cai, Y., Weng, K., Guo, Y., Peng, J., & Zhu, Z. J., (2015). An integrated targeted metabolomic platform for high-throughput metabolite profiling and automated data processing. *Metabolomics, 11*(6), 1575–1586.
54. Li, H., Cai, Y., Guo, Y., Chen, F., & Zhu, Z. J., (2016). MetDIA: Targeted metabolite extraction of multiplexed MS/MS spectra generated by data-independent acquisition. *Anal. Chem., 88*(17), 8757–8764.
55. Chang, H. Y., Chen, C. T., Lih, T. M., Lynn, K. S., Juo, C. G., Hsu, W. L., & Sung, T. Y., (2016). iMet-Q: A user-friendly tool for label-free metabolomics quantitation using dynamic peak-width determination. *PLoS One*, 11, e0146112.
56. Ye, H., Zhu, L., Sun, D., Luo, X., Lu, G., Wang, H., Wang, J., et al., (2016). Nontargeted diagnostic ion network analysis (NINA): Software to streamline the analytical workflow for untargeted characterization of natural medicines. *J. Pharm. Biomed. Anal., 131*, 40–47.
57. Spalding, J. L., Cho, K., Mahieu, N. G., Nikolskiy, I., Llufrio, E. M., Johnson, S. L., & Patti, G. J., (2016). Bar coding MS² spectra for metabolite identification. *Anal. Chem., 88*(5), 2538–2542.
58. Matsuda, F., (2016). Technical challenges in mass spectrometry-based metabolomics. *Mass Spectrom., 5*(2), S0052.
59. Aguilar-Mogas, A., Sales-Pardo, M., Navarro, M., Tautenhahn, R., Guimera, R., & Yanes, O., (2016). *iMet: A Computational Tool for Structural Annotation of Unknown Metabolites from Tandem Mass Spectra*. arXiv Preprint,1607.4122.
60. Zhang, M., Sun, J., & Chen, P., (2015). FlavonQ: An automated data processing tool for profiling flavone and flavonol glycosides with ultra-high-performance liquid chromatography-diode array detection-high resolution accurate mass-mass spectrometry. *Anal. Chem., 87*(19), 9974–9981.
61. Domingo-Almenara, X., Montenegro-Burke, J. R., Benton, H. P., & Siuzdak, G., (2018). Annotation: A computational solution for streamlining metabolomics analysis. *Anal. Chem., 90*(1), 480–489.

62. Fu, H. Y., Guo, X. M., Zhang, Y. M., Song, J. J., Zheng, Q. X., Liu, P. P., Lu, P., et al., (2017). AntDAS: Automatic data analysis strategy for UPLC-QTOF-based nontargeted metabolic profiling analysis. *Anal. Chem., 89*(20), 11083–11090.

63. Pluskal, T., Castillo, S., Villar-Briones, A., & Oresic, M., (2010). MZmine 2: Modular framework for processing, visualizing, and analyzing mass spectrometry-based molecular profile data. *BMC Bioinform., 11*(1), 395.

64. Smith, C. A., Want, E. J., O'Maille, G., Abagyan, R., & Siuzdak, G., (2006). XCMS: Processing mass spectrometry data for metabolite profiling using nonlinear peak alignment, matching, and identification. *Anal. Chem., 78*(3), 779–787.

65. Tsugawa, H., Cajka, T., Kind, T., Ma, Y., Higgins, B., Ikeda, K., Kanazawa, M., et al., (2015). MS-DIAL: Data-independent MS/MS deconvolution for comprehensive metabolome analysis. *Nat. Methods, 12*(6), 523–526.

66. Fu, H. Y., Guo, J. W., Yu, Y. J., Li, H. D., Cui, H. P., Liu, P. P., Wang, B., Wang, S., & Lu, P., (2016). A simple multi-scale Gaussian smoothing-based strategy for automatic chromatographic peak extraction. *J. Chromatogr. A, 1452*, 1–9.

67. Yu, Y. J., Zheng, Q. X., Zhang, Y. M., Zhang, Q., Zhang, Y. Y., Liu, P. P., Lu, P., et al., (2019). Automatic data analysis workflow for ultra-high performance liquid chromatography-high resolution mass spectrometry-based metabolomics. *J. Chromatogr. A, 1585*, 172–181.

68. Liu, P., Zhou, H., Zheng, Q., Lu, P., Yu, Y. J., Cao, P., Chen, W., & Chen, Q., (2019). An automatic UPLC-HRMS data analysis platform for plant metabolomics. *Plant Biotechnol. J., 17*(11), 2038–2040.

69. Tautenhahn, R., Patti, G. J., Rinehart, D., & Siuzdak, G., (2012). XCMS online: A web-based platform to process untargeted metabolomic data. *Anal. Chem., 84*(11), 5035–5039.

70. Mehl, F., Gallart-Ayala, H., Konz, I., Teav, T., Oikonomidi, A., Peyratout, G., Van, D. V. V., et al., (2018). LC-HRMS data as a result of untargeted metabolomic profiling of human cerebrospinal fluid. *Data Brief, 21*, 1358–1362.

## CHAPTER 3

# Dereplication: HRMS in Phytochemical Analysis

SHINTU JUDE and SREERAJ GOPI

*Research and Development (R&D) Center, Plant Lipids (P) Ltd.,*
*Kadayiruppu, Kolenchery, Cochin, Ernakulam, Kerala – 682311, India*

## ABSTRACT

Dereplication is a significant strategy that plays a role in phytochemical investigations. While dealing with bioactive phytochemical entities, a preliminary information on the presence/absence of compounds, their potentials and physicochemical natures provide a versatile cornerstone for further characterization, especially when involve a large number of samples or compounds. A tailor-made workflow can be fabricated for the physicochemical characterization, out of the available technologies, datasets, repositories, and other data processing tools. This chapter deals with the dereplication strategies, with necessary examples. Also, bring some idea of the important significant dereplication techniques among the ever-generated workflows.

## 3.1 INTRODUCTION

The natural products, especially those are originated from plant play a major role in many fields such as food, pharma, beverages, nutraceuticals, etc. In the ancient times, victuals were not a big research material. However, education and technology impelled the human kind not to trust anything blindly and to dig out the facts. Now, it is not enough for him to get the things with effectiveness, they have to be safe also. So he started to screen everything—the bioactives, toxins, coloring ingredients, etc., in his eatables. Moreover, herbal remedies with therapeutic properties and even the food itself came into action than drugs, and many drugs were happened to source from natural products itself. In addition, many of the synthetic drugs have been taken model from natural compounds [1]. In other words, the border between foods-nutraceuticals-drugs became unidentifiable. In this perspective, there are many studies focused to identify the compounds present in the edible naturals, especially in the traditional therapeutic materials, culinary herbs, and spices. The knowledge of significant bioactive compounds present in the plants impelled to conduct more studies on phytochemicals. However, the screening of bioactive components from plant matrix is laborious as it includes a series of lengthy analysis steps, use of costly consumables, and misleading data from the traditional instruments and so on.

Natural matrices are complex mixtures of components and much time and efforts are required for the isolation, identification, and characterization of compounds from them. In new product identification schemes, sometimes these efforts may end up in the rediscovery of known compounds. Much time has been wasted in the area of natural product investigation, for the rediscovery of known compounds. In 1978, when the word 'dereplication' was mentioned for the first time, it didn't create much impact in the related fields [2, 3]. But now, after a few decades, the word act as a synonym for all the screening processes and related techniques for differentiating the already known components from a matrix, by identifying their molecular formula, molecular weight, structure, bioactivity, and taxonomy. Advancements in the science of instrumentation opened the door to the world of detection and identification, and dereplication of complex natural matrices made the identification and characterization of new compounds easier, as it reduces the chances of false-positive results.

In a vast meaning, the term dereplication represents a process flow (Figure 3.1), consisting of:

1. Identification/selection of plant source;
2. Characterization: scheme for extraction and purification, if any, as well as the chromatographic separation and mass spectral data collection of the sample;
3. Bioactivity determination, an optional procedure;
4. Construction/collection of spectral database from the spectrometric characterization of reference compounds;
5. Determinations of compound existence and investigation strategies.

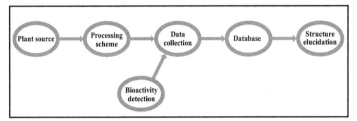

**FIGURE 3.1**    The process flow for ideal dereplication program.

## 3.2  SELECTION CRITERIA

The strategies for dereplication and discovery of bioactive compounds begin with the proper selection of materials and instruments. There are many systems and approaches were introduced, behalf of the same. Traditional medicinal practices rely basically on the natural products and their influences categorized the first approach where, the clinically established beneficial traditional therapeutic drugs are investigated for their characterization. Here, the only unknown thing is the components involved. A prior knowledge of the herbal preparations leads way for sample preparation and the awareness on treating diseases helps in selecting the procedure and makes the task easier. For example, 88 plant extracts, which were established as the traditional medicines for treating snakebites, were screened for their active inhibitors against necrotizing enzymes in snake venoms, and the activity and structural analysis together built up a strong platform for dereplication [4]. Another important study of such kind was the identification of natural fungicides from Ghanian *Uvaria chamae* P. Beauv, a traditional antibiotic [5]. A dereplication workflow was successfully aided in the identification of active compounds and mechanism behind the action of plants from the genus *Rhodiola* L. for the treatment and prevention of acute mountain sickness, used in Tibetan medicine practices [6].

In a second approach, chemotaxonomical/taxonomical data were used for establishing the secondary metabolites of similar nature from related taxa and to derive structure-activity relationships (SAR) thereof. In a strategy applied for the identification of active components of *Spatholobus suberectus*, the fragmentation pathways are tooled for grouping similar compounds and differentiating structural isomers [7].

The third approach is based on the geographic platforms. All the species in a geographical area are evaluated for the biological activities and these results in libraries as well as genetic banks containing a broad and diverse spectrum of bioactive phytochemicals. Many studies from Simirgiotis and team have dealt with the characterization of different endemic species from South America, out of which 5 were handling specifically the species from Paposo Valley, located on the cost of the Atacama Desert [8–12]. Together, these reports form a strong reference for a number of compounds.

The next approach was developed as filler in the voids of all these aforementioned methods. In this approach, the information on the biological activities of plants and the related diseases or assays is intelligently utilized for the isolation and characterization of active compounds, which have not been studied previously. A study, which involved the characterization of phenolic acids (PA) from *Salvia miltiorrhiza*, was intended to investigate the chemical conversion products of PA's and their detection [13]. Another study was stimulated by the fact that, the germination of parasite '*Orobanche cumana*' is induced by the host plant metabolites and only one stimulant had been identified till the time [14]. These studies have executed on the anticipation of the presence of more bioactives in the plants, which could be responsible for the activities, along with/other than the known compounds.

The approach stands independent and takes a higher position, because here, the observation, knowledge, and curiosity of the scientist play the key role behind the discovery of plant bioactives [15].

## 3.3   CHARACTERIZATION PROCEDURES

Many isolation and characterization procedures are outlined for biologically active molecules.

### 3.3.1   UNTARGETED COMPOUNDS SCREENING

The widely used basic characterization strategy is the 'untargeted screening,' which is applied for the identification of major compounds. It is characterized by the complete profiling of all the responding analytes in the sample, which enhances the capabilities of identification of novel compounds. In a non-targeted workflow, the total ion chromatogram

(TIC) is considered for the characterization. In TIC, the major components are selected manually or by program and then each abundant peak are examined for their exact mass spectrum, followed by elemental composition. The possible structural properties are assigned by using references and databases such as similarity analysis (SA), hierarchical clustering analysis (HCA), principle component analysis (PCA), partial least squares-discriminant analysis (PLS-DA), and orthogonal projection to latent structures discriminate analysis (OPLS-DA), etc. In some cases, the accurate mass fingerprinting could serve well for confirmation and quality assurance [16–18]. A similar chemometric approach was used for the differentiation of ginseng roots of different origin and different ages [19]. In another study, a workflow consisting of the data acquisition, RT alignment, statistical analysis, and species identification was achieved and resulted in metabolome fingerprinting [20]. However, this approach fails in many cases such as, if the active compound present in trace amounts, or it delivers a synergic effect with any other components, etc. Another hurdle is the reproducibility of ionization patterns, because each HRMS technique facilitates different ionization technique. The availability of standardized mass spectral databases-home-made or theoretical-can accelerate the process.

### 3.3.2   UNTARGETED PROFILING FOR METABOLOMICS

In accordance with the growth of metabolomics, many more strategies were introduced for the untargeted profiling, for the collection of all possible data of compounds in the sample. With the help of an appropriate informatics tool, it is possible to extract the needed information to be used for different requirements such as chemical fingerprinting (CF), biomarker characterization, quality assurance, etc., even a long time after data collection. Cai Tie and crew have introduced a strategy based on oligosaccharides profiling for the classification of 52 different *Epimedium* herbs by the identification of biomarker compounds [21]. Possibilities of correlating the HRMS chemometric fingerprints with bioactivity were successfully utilized in the investigation of antioxidant anthocyanins from six Chilean berries and phenolic compounds from South American fruits as well as shrubs [22–24].

### 3.3.3   ACTIVITY-GUIDED ISOLATION OF ACTIVE COMPOUNDS

Another important revolution was the activity-guided isolation. It involves the series of extraction and activity determination. The fractions which show the activity is further proceeded for the re-fractionation. This process of fractionation is continued for the active fraction, which will result in a purified bioactive fraction. The systematic monitoring not only allows the online detection of active compounds, but also verifies the route of negative results. The antidiabetic constituents from *Radix Scutellariae* are an illustration of this strategy. The study put forward the potential of, three different bioassays, i.e., aldose reductase, α-glucosidase, and radical scavenging inhibition assays for the activity profiling corresponding to three different pathways of diabetes, which were correlated to the HPLC separation in order to obtain the biochromatograms, and the active compounds were characterized by using a platform of HPLC-HRMS-SPE-NMR [25]. The same platform was used for the characterization of isoflavones from *Azorella madreporica,* antioxidants from *Gomortega keule*, etc., [26, 27]. There are many technologies and methods were presented to enhance the bioactivity guided isolation process, which will be discussed in the following chapters. Being bioassay-guided fractionation as the core, many programs were developed for the purification of compounds, detection of bioactivity of compounds and their synergic effects, chemical profiling, information regarding false positives, etc. However, here, the activity observed in the extract may not appear in the isolated compounds due to many factors such as degradation, chemical changes, loss of synergy, etc.

### 3.3.4   CONSTRUCTION OF NATURAL PRODUCT LIBRARIES

Another important alternative was the construction of natural product libraries. Here, irrespective of the bioactivities, natural compounds of different chemical structures were isolated and used for building up a library, which may then be used for bioactivity characterization. An ideal library can include the information regarding elemental composition, chemical structures, fragmentation pattern, etc., and confirmative evidences are added from reference standards. With the help of HPLC-SPE-NMR, this concept was illustrated by the targeted isolation of compounds from *Hubertia ambavilla* Bory and *Hubertia tomentosa* Bory

(Asteraceae). 17 major peaks obtained in the HPLC separation were selected for further investigation, which have resulted in identification of three new compounds and their subsequent activities. These data were used for the construction of a natural product library of the compounds, allowing a simple and advanced leap towards the new isolated compounds [28]. Another notable work was the generation of spectral library of 252 new psychoactive substances (NPS) of different classes and the preparation of a database comprising chemical and structural details of 875 NPS [29].

## 3.4   DIFFERENT TECHNICAL AND INSTRUMENTAL CONCERNS

In order to build a characterization platform, it is necessary to obtain the isolated recognition of compounds even in small amounts, exact mass, elemental composition, fragmentation patterns and sometimes, special arrangements also. Therefore, the analytical platforms corresponding to the separation, detection, processing, and structural characterization together form dereplication modules. GC and other thermal techniques solely depend on the thermal properties of the extracts. So, in many cases, they are not used for the separation and ionization for the natural product characterization. Besides, they do not support the screening of broad spectra of compounds with different properties and structures. Most of the dereplication studies have followed a trial and error system of methods, i.e., many modes of hyphenation have executed for a single sample, which produces various results. By analyzing these results, a most reliable combination of instruments and techniques is assigned. For example, in dereplication of *Salvia miltiorrhiza,* many separation aids such as reverse phase liquid chromatography (RPLC), UHPLC, electrophoresis, etc., separation modes such as ion exchange, size exclusion, hydrophilic interaction, affinity interaction, normal phase, reversed phase (RP), etc., and detection techniques such as DAD, MS, HRMS, etc., were employed in different studies, in search of better results. An offline two-dimensional (2D) liquid chromatography-a combination of two orthogonal LC systems-aided with hydrophilic interaction chromatography column and RP column have been proved to be enhancing selectivity and peak capacity [13]. In addition, the hybridization of IT-ToF-MS enables the highly accurate measurements of the ions.

## 3.5  HRMS AND DEREPLICATION

Development of high-resolution mass spectrometry has added color to the picture of dereplication. It could allow the direct identification of structural formula of compounds present in the biofluids and tissues in a single run. Many researchers have proved HRMS to be a powerful technique for full structural analysis of constituents directly from crude extracts, without further purifications. Dereplication involves the rapid identification and sometimes quantification of known compounds, irrespective of their chemical classes and hence accelerates the new compounds discovery processes. It forms the primary step of many analytical strategies such as chemical profiling, bioactivity-guided purifications, fingerprinting of extracts, etc., and became an inevitable segment in many fields such as genomics, metabolomics, taxonomic identification, biochemistry, pharmacology, etc. The potential of combining high-speed counter-current chromatography (HSCCC) and HPLC-UV-HRMS$^n$ methodologies was demonstrated in the screening of the non-volatile chemical composition of *Lippia origanoides*. The complexity of matrix was overwhelmed by the specific hyphenation and found useful in characterization of the flavonoid rich specimen [30]. Similarly, the interspecific diversity of active compounds present in different organs of *Moringa oleifera* was assessed across most of the available species, as part of a study using HPLC separations and mass spectrometric and NMR data [31]. Even a simple open column chromatography prior to the HPLC separation, could exert a multifold synergic purification effect for active components from *Azadirachta indica* and *Melia azedarach* [32].

## 3.6  DATA INTERPRETATION

HRMS provides a huge data, containing information of different manners. Basically, HRMS provides prime information regarding the exact molecular mass. In the advanced forms, it is possible to obtain the elemental compositions to accuracy for mass 0.0001 amu, even in very low concentrations. Structural specificity is linked with molecular mass, and the information on exact mass is the prime data for structure elucidation. According to the dereplication functions, mode of detector readouts can be used in many different strategies, so as to get the results as per the requirements. If the purpose is a qualitative analysis, the primary intention is to verify the presence or absence of a compound or a number of compounds by incorporating the data with databases or references. Hence, all the available data formats are used for it. For compound characterization, the patterns are selected such that, the structure related data of the analyte to be characterized is enhanced and used for elucidation. Quantitative mode of analyzes require the information related to the both, presence and content of sample, with respect to the reference [33]. In any mode of dereplication, MS-NMR instrument hyphenation furnishes a complete detection platform, being two complementary techniques among the array of detectors. NMR delivers the structure related information even including the arrangement of functional groups as well as the regioisomerism. The hyphenation precisely provides an orthogonal data, which simply increases the confidence by multifold for the annotation of compound structure. Mass spectral data, along with NMR information allows the connections and the identification of even stereochemistry of compounds. Multi-hyphenated systems leave an ocean of information behind. Utilizing these data in proper way results in recognition of the activity and finally the compounds themselves.

Analytical instruments play only the first part of detection in dereplication; the rest depends on the processing of the observed data and hence compound databases come into the platform. Analytical instruments, especially HRMS acquire a huge amount of data, and it needs to be processed with proper databases, processing methodologies or reference approaches, in order to derive meaningful results. Various modes of hardware and software were introduced for the data interpretation according to the number and nature of instruments included in the dereplication methodologies. Comparing these experimental data with, references is a standard dereplication procedure. Here, we meet with two terms-databases and data mining. Databases are organized and compiled collection of information, while data mining represents the processing of raw data to obtain meaningful results. Basic facts enclosed with databases are the physicochemical properties of bioactive components, along with their structure, origin, possible methods of isolation, spectrum of activity, etc. Strategies of dereplication based on either identification of compounds or classification into groups are more popular.

### 3.6.1 DATABASES

Quality and proper selection of the databases are important parameters in dereplication. Databases are prepared, formatted, and categorized on basis of many parameters. We can see some examples.

STN, an online database service introduced databases in chemical abstracts service (CAS), in two formats-'CA plus,' a bibliographic database of chemical information and its companion file 'CAS registry,' which is a chemical substance information database [34]. The information is gathered from authorized documents such as patent publications, journals, dissertations, books, etc. Chapman and Hall's 'Dictionary of Natural Products' is another repository of natural chemicals and their properties [35]. 'Bioactive Natural Product Database' presented natural compounds along with their biological activities [36]. 'Natural product activity and species source' (NPASS) is a freely accessible database, which included the experimental activity values and species sources as well [37]. An open platform for compound structures was introduced as DEREP-NP, which consists of the structures of natural products of both plant and animal origin [38]. In the database named 'SuperToxic,' the compounds from both natural and artificial origin, which were proven as toxic, are included with the description on their structural, chemical, and functional properties as well as toxicology [39]. 'Supernatural' is another database with timely improved versions, which provides the structural and physicochemical properties along with the predicted toxicity of bioactive compounds [40]. 'NAPROC-13' is considered as a complementary tool for 'Supernatural,' as it contains the identification parameters and stereo specificity of natural products from phytochemical studies [41]. 'Therapeutic Target Database,' commonly known as TTD provides information of efficacy targets and their corresponding drugs with detailed information on target validation, SARs, clinical, and pre-clinical trials data, etc., [42]. 3DMET is a database of 3D structures of natural compounds, which consists of a self-checking system and two modes for verifying the 3D structures. PlantMAT allows the prediction of plant natural products, enabling dereplication, leaving an opportunity for novel structure discovery [43].

Databases organized with specificity on their discipline are helpful in specific drug development. Cardiovascular disease herbal database (CVDHD) consists of the identification information, molecular properties, and docking results of compounds from medicinal plants for cardiovascular-related diseases [44]. Naturally occurring plant-based anticancerous compound-activity-target database (NPACT) provides the bioactivities of anti-cancer natural compounds along with the related cancer cells and their molecular targets [45].

Some databases have arranged the data based on the geographic region of their experiments. NuBBE$_{DB}$ is an example, which is a database of compounds of Brazilian biodiversity [46]. There are many notable works established for traditional Chinese medicines (TCMs). 'TCM Database@ Taiwan' facilitates virtual screening by providing 3D compound structures of TCM [47]. The 'TCM integrative database-TCMID' is developed based on six modules-prescription, herb, ingredients, disease, target, and drugs, whereas 'Chinese ethnic minority traditional drug database-CEMTDD' is composed of modules-plants, metabolites, active components, indications, targeted proteins, diseases, and mechanisms, along with an access towards herb-compound-disease-target network of mechanism [48, 49]. 'HIT: Herb Ingredients' targets' contains a curated database of molecular targets, their experimental conditions, and observed bioactivity against more than 1300 Chinese herbs [50]. The bioactivity and 3D structures of African medicinal plants were included in AfroDb [51]. In the same way, 3D structure, source, activities, and properties of the compounds from medicinal plants of Cameroonian flora were presented for virtual screening in CamMedNP [52]. A distinct platform for the marine natural product research was provided by MarinLit [53].

## 3.7 DATA-MINING TOOLS

Many computer-based tools were developed so far, useful in peak selection, ion extraction, organizing the data, etc., by utilizing molecular structures, fragmentation patterns, bioactivity, etc., and are generally known as data-mining tools [54]. One interesting advantage for a few techniques among them is the possibility of handling data from different instruments. Some of the tools are provided by the instrument facilitators such as waters (MarkerLynx, MassLynx), Agilent (MassHunter), etc., whereas some others are available publically, such as XCMS, MZmine, etc., [55]. There are data mining tools available which make use of the substructures (e.g., MetFusion), or fragmentation patterns (e.g., ISIS, and FTBLAST) for chemical characterization [56–58]. FingerID deals with the structural parameters and ends up in molecular structures [59]. Another important tool is molecular networking, where, the fragmentation patterns of structurally similar molecular species are correlated and used for the detection of related compounds as well as the investigation of molecular interactions [60]. Even the simpler workflows, which depend on the direct MS infusion or MS²precursor data, were proved to be effective in successful dereplication. An earlier revolution in this field was the development of automated mass spectral deconvolution and identification system (AMDIS), which was tooled for the extraction of pure compound spectra from the obtained real chromatogram [61].

## 3.8    DEREPLICATION STUDIES

Dereplication studies are of two types, while considering the screening pattern. Many works identify new workflows and use them directly for the dereplication purpose and/or act as templates for future studies. At the same time, some of the studies utilize previously aligned workflows for rapid recognition of the known compounds.

### 3.8.1   *DEVELOPMENT OF DEREPLICATION STRATEGIES FOR FUTURE STUDIES*

There is a number of dereplication strategies has been introduced on the HRMS platform to be used with different manners of actions and applications, which were demonstrated with many plants. Let us have a look on some relevant and well-documented studies to familiarize with the technologies, instruments, and workflows, which are involved in the path finding dereplication strategies. The genus *Rauwolfia* was studied extensively for their phytochemical characterization by many instruments including HRMS. They have been experimented under a number of different procedures, by employing the possibilities of ambient ionization, multistage mass analysis, fragmentation, comparison with references, etc. Comprehensive investigation reports are available on the phytochemical components from different organs of the species [62]. Boukhris and his team introduced an on-line dereplication strategy for getting the structural information of phenolic constituents of *Anvillea radiate*. By proceeding with a dual extraction, dual characterization platform, even the minor, and less abundant phenolic compounds were also characterized selectively. Besides, the study provided profiles of each organ separately [63]. Another workflow was developed for the identification of active compounds behind the diabetes mellitus prevention ability of Walnut leaf. The workflow consisted of compound identification by HRMS data, target prediction by chemical similarities and databases, analysis of component-disease target interaction network and confirmation by molecular docking analysis. The presented workflow was demonstrated with the successful recognition of 38 hypoglycemic components out of the identified 130 components [64]. Approaches consisting of ambient ionization in the procedure are comparatively easy to execute and rapid. An illustration is given with DESI in combination with literature or reference standard data, for the detection of alkaloids from different organs of three different plants. Results were verified by comparing with those from conventional methods [65].

### 3.8.2 ANALYZES USING ALREADY BUILT PLATFORMS

Dereplication of compounds from the roots of two species from *Lamiaceae* family was carried out by a previously established platform. The method consisted of ion trap (IT) mass analyzer and orbitrap mass spectrometer (MS) as two complementary analysis systems. Along, the procedure comprised the collection of data regarding exact molecular masses as well as the fragmentation pattern in both positive and negative modes, comparison of mass spectra and retention times (RTs) to those of standard compounds, data processing by utilizing online available databases and literature data, which have ended up in the identification of 39 compounds from both of the samples. Along with the primary dereplication protocols, a secondary comparison was also made for these two closely related species, using the data of dereplication and antioxidant properties [13]. In another study, DART-HRMS data acquisition, along with multivariate statistical analysis together form a platform for identification and comparison of compounds and thereby the origin determination of heroin [66]. In a similar way, the data acquired by UHPLC-HRMS-SPE-NMR was compared with databases and preceded for multivariate analysis to obtain beneficial information about two species of genus *Phyllanthus* L. [67]. An in-house developed library of reported chemical compounds from Apocynaceae family was used as the dereplication platform for the secondary metabolites of *Tabernaemontana catharinensis* leaves [68].

## 3.9  EFFECTS

The major outcome of dereplication results in the hard-less determination of further procedures in compound characterization. Especially, the removal of Pan Assay Interference Compounds (PAINS)—the interfering compounds during the bioassays-prioritize the process flow in proper channel [69]. Along, the information gathered by different instruments brings light on the different properties of the extract, such as spectroscopic, structural, and biological activities.

## 3.10  SIDE-EFFECTS

As seen above, there are many modes, techniques, technologies, methodologies, etc., are introduced, in relation with dereplication. However, if looking apart from the mainstream, there are strategies developed by the same principles for many other applications. Acquisition of high-resolution mass spectra and the data interpretation by pattern recognition platforms as well as databases along with the statistical analyzing tools provides a promising way for elucidating mechanism of therapeutic action of many herbal formulations in animals. Timely variations in the biomarker fingerprint can be the signs of health status and identification as well as correlation of them with references forms the aid for mechanism elucidation. For example, the mechanism of action of *Corydalis yanhusuo* alkaloid on gastric ulcer was elucidated by dissecting the changes of metabolite profiling. The mechanism of action was presented, including the biomarkers and pathways [70]. Likewise, the mechanism of action of *Suanzaoren* decoction (SZRD) on treatment of insomnia, *jieduquyuziyin* prescription on systemic lupus erythematosus, *Yinchenhao* on liver diseases, *Trans*-crocin 4 on Alzheimer's disease (AD) and Dazhu Hongjingtian on acute mountain sickness were elucidated [71–74]. In the same way, stress and reactions of plants also presented. A notable finding in this kind of biochemical pathway elucidation was the demonstration of mechanism of nanotoxicity caused by CuO NPs exposure in plants. The effected pathways were identified to be altered by the exposure and the specific metabolomic changes prove the defense response effects [20].

An extension of the mechanism elucidation can appear as the toxicity evaluation. Under conditions of toxicity, the metabolic profiling represents the pathological patterns and analysis of which provides significant results on the toxicity effects of the drugs. Examinations of the biochemical compositions enable the characterization of toxicity, differentiation of its patterns with respect to many factors such as sources and related pathways [75]. Aristolochic Acid-Induced Nephrotoxicity in Rats, Chuanwu induced cardiac and neural toxicity in rats, etc., were explained by biomarker characterization approach [75, 76]. An interesting study among them investigated the effects of mode of drug administration in the toxicity effects, apart from a mere explanation of the mechanism [77]. Furthermore, the dosage and tenure of administration play roles in toxicity as well [78]. A new stream of approach, based on the network toxicology and mass spectral data was introduced for deriving the mechanism of action of toxicity [79].

## 3.11  PITFALLS

One of the major disadvantage faced by the HRMS based dereplication is the variabilities in raw dataset acquired by different HRMS analyzers in terms of ionization methods, which make it hard to create a universal database [80]. HRMS-SPE-NMR is considered as a potential tool for dereplication. However, it is reported that, there is differences in mass recovery among different SPE cartridges with respect to species, making the appropriate selection of tools a major step in the process [68]. Expectation for an unpredicted activity is another factor to be followed in dereplication, which anticipates the known unknowns-compounds with known basic structure, but different chemical moieties.

## 3.12  CONCLUSION

Phytochemical investigation includes the characterization of secondary metabolites, which leads way towards bioactive drugs and formulations. Dereplication enables the identification of known metabolites, thereby reduces possible false-positive results. It also enables structural identification of known metabolites present in the complex matrix in a single run. The procedure became advanced with the development of HRMS instruments, allowing the direct detection of molecular formulas of the compounds of interest. Thus it forms a strong guide for phytochemical characterization.

## KEYWORDS

- Alzheimer's disease
- cardiovascular disease herbal database
- data mining
- databases
- dereplication
- effect-directed analysis

## REFERENCES

1. Newman, D. J., & Cragg, G. M., (2012). Natural products as sources of new drugs over the 30 years from 1981 to 2010. *J. Nat. Prod., 75*(3), 311–335.

2. Hanka, L. J., Kuentzel, S. L., Martin, D. G., Wiley, P. F., & Neil, G. L., (1978). Detection and assay of antitumor antibiotics. In:*Antitumor Antibiotics* (pp. 69–76). Springer, Berlin, Heidelberg.

3. Gaudêncio, S. P., & Pereiraa, F., (2015). Dereplication: Racing to speed up the natural products discovery process. *Nat. Prod. Rep., 32*(6), 779–810.

4. Liu, Y., Staerk, D., Nielsen, M. N., Nyberg, N., & Jäger, A. K., (2015). High-resolution hyaluronidase inhibition profiling combined with HPLC-HRMS-SPE-NMR for identification of anti-necrosis constituents in Chinese plants used to treat snakebite. *Phytochemistry,119*, 62–69.

5. Kongstad, K. T., Wubshet, S. G., Kjellerup, L., Winther, A. M. L., & Staerk, D., (2015). Fungal plasma membrane H+-ATPase inhibitory activity of o-hydroxy benzylated flavanones and chalcones from *Uvaria chamae* P. beauv. *Fitoterapia, 105*, 102–106.

6. Ou, C., Geng, T., Wang, J., Gao, X., Chen, X., Luo, X., Tong, X., et al., (2020). Systematically investigating the pharmacological mechanism of dazhu hongjingtian in the prevention and treatment of acute mountain sickness by integrating UPLC/Q-TOF-MS/MS analysis and network pharmacology. *J. Pharm. Biomed. Anal., 179*, 113028.

7. Zhang, Z., Bo, T., Bai, Y., Ye, M., An, R., Cheng, F., & Liu, H., (2015). Quadrupole time-of-flight mass spectrometry as a powerful tool for demystifying traditional Chinese medicine. *TrAC Trends Anal. Chem., 72*, 169–180.

8. Simirgiotis, M. J., Quispe, C., Mocan, A., Villatoro, J. M., Areche, C., Bórquez, J., Sepúlveda, B., & Echiburu-Chau, C., (2017). UHPLC high resolution orbitrap metabolomic fingerprinting of the unique species *Ophryosporus triangularis* Meyen from the Atacama Desert, Northern Chile. *Brazilian Journal of Pharmacognosy, 27*(2), 179–187.

9. Simirgiotis, M. J., Quispe, C., Bórquez, J., Schmeda-Hirschmann, G., Avendaño, M., Sepúlveda, B., & Winterhalter, P., (2016). Fast high-resolution orbitrap MS fingerprinting of the resin of *Heliotropiumtaltalense* phil. from the Atacama Desert. *Ind. Crops Prod., 85*, 159–166.

10. Simirgiotis, M. J., Benites, J., Areche, C., & Sepúlveda, B., (2015). Antioxidant capacities and analysis of phenolic compounds in three endemic Nolana species by HPLC-PDA-ESI-MS. *Molecules, 20*(6), 11490–11507.

11. Simirgiotis, M. J., Quispe, C., Bórquez, J., Mocan, A., & Sepúlveda, B., (2016). High resolution metabolite fingerprinting of the resin of *Baccharistola* Phil. from the Atacama Desert and its antioxidant capacities. *Ind. Crops Prod., 94*, 368–375.

12. Simirgiotis, M. J., Bórquez, J., Neves-Vieira, M., Brito, I., Alfaro-Lira, S., Winterhalter, P., Echiburú-Chau, C., et al., (2015). Fast isolation of cytotoxic compounds from the native Chilean species *Gypothamniumpinifolium* Phil. collected in the Atacama Desert, northern Chile. *Ind. Crops Prod., 76*, 69–76.

13. Ożarowski, M., Piasecka, A., Gryszczyńska, A., Sawikowska, A., Pietrowiak, A., Opala, B., Mikołajczak, P. Ł., et al., (2017). Determination of phenolic compounds and diterpenes in roots of *Salviamiltiorrhiza* and *Salvia przewalskii* by two LC-MS tools: Multi-stage and high-resolution tandem mass spectrometry with assessment of antioxidant capacity. *Phytochem. Lett., 20*, 331–338.

14. Raupp, F. M., & Spring, O., (2013). New sesquiterpene lactones from sunflower root exudate as germination stimulants for *Orobanchecumana. J. Agr. Food Chem., 61*(44), 10481–10487.

15. Cordell, G. A., Beecher, C. W., Kinghorn, A. D., Pezzuto, J. M., Constant, H. L., Chai, H. B., Fang, L., et al., (1996). The dereplication of plant-derived natural products. In: Rahman, A. U., (ed.), *Studies in Natural Products Chemistry* (Vol. 19, pp. 749–791). Elsevier.

16. Liang, X. M., Jin, Y., Wang, Y. P., Jin, G. W., Fu, Q., & Xiao, Y. S., (2009). Qualitative and quantitative analysis in quality control of traditional Chinese medicines. *J. Chromatogr. A, 1216*(11), 2033–2044.

17. Fan, C., Deng, J., Yang, Y., Liu, J., Wang, Y., Zhang, X., Fai, K., et al., (2013). Multi-ingredients determination and fingerprint analysis of leaves from *Ilex latifolia* using ultra-performance liquid chromatography coupled with quadrupole time-of-flight mass spectrometry. *J. Pharm. Biomed. Anal., 84*, 20–29.

18. Li, Y., Zhang, T., Zhang, X., Xu, H., & Liu, C., (2010). Chemical fingerprint analysis of *Phellodendri amurensis* cortex by ultra-performance LC/Q-TOF-MS methods combined with chemometrics. *J. Sep. Sci., 33*(21), 3347–3353.

19. Lee, D. Y., Cho, J. G., Bang, M. H., Han, M. W., Lee, M. H., Yang, D. C., & Baek, N. I., (2011). Discrimination of Korean ginseng (*Panax ginseng*) roots using rapid resolution LC-QTOF/MS combined by multivariate statistical analysis. *Food Sci. Biotechnol., 20*(4), 1119.

20. Soria, N. G. C., Bisson, M. A., Atilla-Gokcumen, G. E., & Aga, D. S., (2019). High-resolution mass spectrometry-based metabolomics reveal the disruption of Jasmonic pathway in Arabidopsis thaliana upon copper oxide nanoparticle exposure. *Sci. Total Environ., 693*, 133443.

21. Tie, C., Hu, T., Guo, B., & Zhang, J., (2015). Novel strategy for herbal species classification based on UPLC-HRMS oligosaccharide profiling. *J. Pharm. Biomed. Anal., 111*, 14–20.

22. Ramirez, J. E., Zambrano, R., Sepúlveda, B., Kennelly, E. J., & Simirgiotis, M. J., (2015). Anthocyanins and antioxidant capacities of six Chilean berries by HPLC–HR-ESI-ToF-MS. *Food Chem., 176*, 106–114.

23. Simirgiotis, M. J., Bórquez, J., & Schmeda-Hirschmann, G., (2013). Antioxidant capacity, polyphenolic content and tandem HPLC-DAD-ESI/MS profiling of phenolic compounds from the South American berries *Luma apiculata* and *L. Chequén*. *Food Chem., 139*(1–4), 289–299.

24. Echiburu-Chau, C., Pastén, L., Parra, C., Bórquez, J., Mocan, A., & Simirgiotis, M. J., (2017). High resolution UHPLC-MS characterization and isolation of main compounds from the antioxidant medicinal plant *Parastrephia lucida* (Meyen). *Saudi Pharm. J., 25*(7), 1032–1039.

25. Tahtah, Y., Kongstad, K. T., Wubshet, S. G., Nyberg, N. T., Jønsson, L. H., Jäger, A. K., Qinglei, S., & Staerk, D., (2015). Triple aldose reductase/α-glucosidase/radical scavenging high-resolution profiling combined with high-performance liquid chromatography-high-resolution mass spectrometry-solid-phase extraction-nuclear magnetic resonance spectroscopy for identification of antidiabetic constituents in crude extract of *Radix scutellariae*. *J. Chromatogr. A, 1408*, 125–132.

26. Bórquez, J., Kennelly, E. J., & Simirgiotis, M. J., (2013). Activity guided isolation of isoflavones and hyphenated HPLC-PDA-ESI-ToF-MS metabolome profiling of *Azorellamadreporica* Clos. from northern Chile. *Food Res. Int., 52*(1), 288–297.

27. Simirgiotis, M. J., Ramirez, J. E., Hirschmann, G. S., & Kennelly, E. J., (2013). Bioactive coumarins and HPLC-PDA-ESI-ToF-MS metabolic profiling of edible queule fruits (*Gomortega keule*), an endangered endemic Chilean species. *Food Res. Int., 54*(1), 532–543.

28. Sprogøe, K., Stærk, D., Jäger, A. K., Adsersen, A., Hansen, S. H., Witt, M., Landbo, A. K. R., et al., (2007). Targeted natural product isolation guided by HPLC-SPE-NMR: Constituents of hubertia species. *J. Nat. Prod., 70*(9), 1472–1477.

29. Seither, J. Z., Hindle, R., Arroyo-Mora, L. E., & DeCaprio, A. P., (2018). Systematic analysis of novel psychoactive substances: I. Development of a compound database and HRMS spectral library. *Forensic Chem., 9*, 12–20.

30. Leitão, S. G., Leitão, G. G., Vicco, D. K., Pereira, J. P. B., De Morais, S. G., Oliveira, D. R., Celano, R., et al., (2017). Counter-current chromatography with off-line detection by ultra-high performance liquid chromatography/high resolution mass spectrometry in the study of the phenolic profile of *Lippia origanoides*. *J. Chromatogr. A., 1520*, 83–90.

31. Fahey, J. W., Olson, M. E., Stephenson, K. K., Wade, K. L., Chodur, G. M., Odee, D., Nouman, W., et al., (2018). The diversity of chemoprotective glucosinolates in *Moringaceae* (*Moringa spp.*). *Sci. Rep., 8*(1), 1–14.

32. Caboni, P., Ntalli, N. G., Bueno, C. E., & Alche, L. E., (2012). Isolation and chemical characterization of components with biological activity extracted from *Azadirachta indica* and *Melia azedarach*. In: Patil, B. S., Jayaprakasha, G. K., Murthy, K. N. C., & Seeram, N. P., (eds.), *Emerging Trends in Dietary Components for Preventing and Combating Disease* (pp. 51–77). American Chemical Society.

33. Seger, C., Sturm, S., & Stuppner, H., (2013). Mass spectrometry and NMR spectroscopy: Modern high-end detectors for high-resolution separation techniques-state of the art in natural product HPLC-MS, HPLC-NMR, and CE-MS hyphenations. *Nat. Prod. Rep., 30*(7), 970–987.

34. Williams, J., & Ebe, T., (1997). STN easy: Point-and-click patent searching on the World Wide Web. *World Patent Information, 19*(3), 161–166.

35. Running, W., (1993). Computer software reviews. Chapman and Hall dictionary of natural products on CD-ROM. *J. Chem. Inf. Comput. Sci., 33*(6), 934–935.

36. Berdy, J., & Kertesz, M., (1989) *Chemical Information*. Springer: Berlin, Heidelberg.

37. Zeng, X., Zhang, P., He, W., Qin, C., Chen, S., Tao, L., Wang, Y., et al., (2018). NPASS: Natural product activity and species source database for natural product research, discovery, and tool development. *Nucleic Acids Res., 46*(D1), D1217–D1222.

38. Zani, C. L., & Carroll, A. R., (2017). Database for rapid dereplication of known natural products using data from MS and fast NMR experiments. *J. Nat. Prod., 80*(6), 1758–1766.

39. Schmidt, U., Struck, S., Gruening, B., Hossbach, J., Jaeger, I. S., Parol, R., Lindequist, U., et al., (2009). Supertoxic: A comprehensive database of toxic compounds. *Nucleic Acids Res., 37*(1), 295–299.

40. Dunkel, M., Fullbeck, M., Neumann, S., & Preissner, R., (2006). Supernatural: A searchable database of available natural compounds. *Nucleic Acids Res., 34*(1), 678–683.

41. López-Pérez, J. L., Therón, R., Del Olmo, E., & Díaz, D., (2007). NAPROC-13: A database for the dereplication of natural product mixtures in bioassay-guided protocols. *Bioinformatics, 23*(23), 3256–3257.

42. Wang, Y., Zhang, S., Li, F., Zhou, Y., Zhang, Y., Wang, Z., Zhang, R., et al., (2020). Therapeutic target database 2020: Enriched resource for facilitating research and early development of targeted therapeutics. *Nucleic Acids Res., 48*(D1), D1031–D1041.

43. Qiu, F., Fine, D. D., Wherritt, D. J., Lei, Z., & Sumner, L. W., (2016). PlantMAT: A metabolomics tool for predicting the specialized metabolic potential of a system and for large-scale metabolite identifications. *Anal. Chem., 88*(23), 11373–11383.

44. Gu, J., Gui, Y., Chen, L., Yuan, G., & Xu, X., (2013). CVHD: A cardiovascular disease herbal database for drug discovery and network pharmacology. *J. Cheminformatics, 5*(1), 51.

45. Wang, Y., Zhang, S., Li, F., Zhou, Y., Zhang, Y., Wang, Z., Zhang, R., et al., (2020). Therapeutic target database 2020: Enriched resource for facilitating research and early development of targeted therapeutics. *Nucleic Acids Res., 48*(1), 1031–1041.

46. Pilon, A. C., Valli, M., Dametto, A. C., Pinto, M. E. F., Freire, R. T., Castro-Gamboa, I., Andricopulo, A. D., & Bolzani, V. S., (2017). NuBBE DB: An updated database to uncover chemical and biological information from Brazilian biodiversity. *Sci. Rep., 7*(1), 1–12.

47. Chen, C. Y. C., (2011). TCM Database@ Taiwan: The world's largest traditional Chinese medicine database for drug screening in silico. *PloS One, 6*(1).

48. Xue, R., Fang, Z., Zhang, M., Yi, Z., Wen, C., & Shi, T., (2012). TCMID: Traditional Chinese medicine integrative database for herb molecular mechanism analysis. *Nucleic Acids Research, 41*(D1), D1089-D1095.

49. Huang, J., & Wang, J. H., (2014). CEMTDD: Chinese ethnic minority traditional drug database. *Apoptosis, 19*(9), 1419–1420.
50. Ye, H., Ye, L., Kang, H., Zhang, D., Tao, L., Tang, K., Liu, X., et al., (2010). HIT: Linking herbal active ingredients to targets. *Nucleic Acids Res., 39*(1), D1055–D1059.
51. Ntie-Kang, F., Zofou, D., Babiaka, S. B., Meudom, R., Scharfe, M., Lifongo, L. L., Mbah, J. A., et al., (2013). AfroDb: A select highly potent and diverse natural product library from African medicinal plants. *PLoS One,* 8(10).
52. Ntie-Kang, F., Mbah, J. A., Mbaze, L. M. A., Lifongo, L. L., Scharfe, M., Hanna, J. N., Cho-Ngwa, F., et al., (2013). CamMedNP: Building the Cameroonian 3D structural natural products database for virtual screening. *BMC Complem. Altern. M, 13*(1), 1–10.
53. Munro, M. H. G., & Blunt, J. W., (2007). *MARINLIT, a Database of the Marine Natural Products Literature* (p. 12). Department of Chemistry, University of Canterbury, Christchurch, New Zealand, vpc.
54. Hubert, J., Nuzillard, J. M., & Renault, J. H., (2017). Dereplication strategies in natural product research: How many tools and methodologies behind the same concept? *Phytochem. Rev., 16*(1), 55–95.
55. Gürdeniz, G., Kristensen, M., Skov, T., & Dragsted, L. O., (2012). The effect of LC-MS data preprocessing methods on the selection of plasma biomarkers in fed vs. fasted rats. *Metabolites, 2*(1), 77–99.
56. Gerlich, M., & Neumann, S., (2013). MetFusion: Integration of compound identification strategies. *Journal of Mass Spectrometry, 48*(3), 291–298.
57. Earnhardt, J. M., Thompson, S. D., & Willis, K., (1995). ISIS database: An evaluation of records essential for captive management. *Zoo Biology, 14*(6), 493–508.
58. Ludwig, M., Hufsky, F., Elshamy, S., & Böcker, S., (2012). Finding characteristic substructures for metabolite classes. In: Böcker, S., Hufsky, F., Scheubert, K., Schleicher, J., & Schuster, S., (ed.),*German Conference on Bioinformatics* (p. 23). Leibniz-Zentrum für Informatik:.
59. Dührkop, K., Shen, H., Meusel, M., Rousu, J., & Böcker, S., (2015). Searching molecular structure databases with tandem mass spectra using CSI: FingerID. *Proc. Natl. Acad. Sci., 112*(41), 12580–12585.
60. Yang, J. Y., Sanchez, L. M., Rath, C. M., Liu, X., Boudreau, P. D., Bruns, N., Glukhov, E., et al., (2013). Molecular networking as a dereplication strategy. *J. Nat. Prod., 76*(9), 1686–1699.
61. Vey, S., & Voigt, A., (2007). AMDiS: Adaptive multidimensional simulations. *Computing and Visualization in Science, 10*(1), 57–67.
62. Kumar, S., Singh, A., Bajpai, V., Srivastava, M., Singh, B. P., & Kumar, B., (2016). Structural characterization of monoterpene indole alkaloids in ethanolic extracts of Rauwolfia species by liquid chromatography with quadrupole time-of-flight mass spectrometry. *J. Pharm. Anal., 6*(6), 363–373.
63. Boukhris, M. A., Destandau, É., El Hakmaoui, A., El Rhaffari, L., & Elfakir, C., (2016). A dereplication strategy for the identification of new phenolic compounds from *Anvillearadiata* (Coss. & Durieu). *Comptes Rendus Chimie, 19*(9), 1124–1132.
64. Liu, R., Su, C., Xu, Y., Shang, K., Sun, K., Li, C., & Lu, J., (2020). Identifying potential active components of walnut leaf that acton diabetes mellitus through

integration of UHPLC-Q-orbitrap HRMS and network pharmacology analysis. *J. Ethnopharmacol., 112*659.

65. Talaty, N., Takáts, Z., & Cooks, R. G., (2005). Rapid in situ detection of alkaloids in plant tissue under ambient conditions using desorption electrospray ionization. *Analyst, 130*(12), 1624–1633.

66. Cui, X., Lian, R., Chen, J., Ni, C., Liang, C., Chen, G., & Zhang, Y., (2019). Source identification of heroin by rapid detection of organic impurities using direct analysis in real time with high-resolution mass spectrometry and multivariate statistical analysis. *Microchem. J., 147*, 121–126.

67. Silva, M. F. S., Silva, L. M. A., Quintela, A. L., Dos, S. A. G., Silva, F. A. N., Fátima, D. C. E., Alves, F. E. G., et al., (2019). UPLC-HRMS and NMR applied in the evaluation of solid-phase extraction methods as a rational strategy of dereplication of *Phyllanthus* spp. aiming at the discovery of cytotoxic metabolites. *J. Chromatogr. B, 1120*, 51–61.

68. Camponogara, C., Casoti, R., Brusco, I., Piana, M., Boligon, A. A., Cabrini, D. A., Trevisan, G., et al., (2019). *Tabernaemontanacatharinensis* leaves exhibit topical anti-inflammatory activity without causing toxicity. *J. Ethnopharmacol., 231*, 205–216.

69. Dahlin, J. L., Nissink, J. W. M., Strasser, J. M., Francis, S., Higgins, L., Zhou, H., Zhang, Z., & Walters, M. A., (2015). PAINS in the assay: Chemical mechanisms of assay interference and promiscuous enzymatic inhibition observed during a sulfhydryl-scavenging HTS. *J. Med. Chem., 58*(5), 2091–2113.

70. Tianjiao, L., Shuai, W., Xiansheng, M., Yongrui, B., Shanshan, G., Bo, L., Lu, C., et al., (2014). Metabolomics coupled with multivariate data and pathway analysis on potential biomarkers in gastric ulcer and intervention effects of *Corydalisyanhusuo* alkaloid. *PLoS One., 9*(1),e82499.

71. Yang, B., Zhang, A., Sun, H., Dong, W., Yan, G., Li, T., & Wang, X., (2012). Metabolomic study of insomnia and intervention effects of suanzaoren decoction using ultra-performance liquid-chromatography/electrospray-ionization synapt high-definition mass spectrometry. *J. Pharm. Biomed. Anal., 58*, 113–124.

72. Ding, X., Hu, J., Wen, C., Ding, Z., Yao, L., & Fan, Y., (2014). Rapid resolution liquid chromatography coupled with quadrupole time-of-flight mass spectrometry-based metabolomics approach to study the effects of Jieduquyuziyin prescription on systemic lupus erythematosus. *PLoS One, 9*(2),e88223.

73. Sun, H., Zhang, A. H., Zou, D. X., Sun, W. J., Wu, X. H., & Wang, X. J., (2014). Metabolomics coupled with pattern recognition and pathway analysis on potential biomarkers in liver injury and hepatoprotective effects of yinchenhao. *Appl. Biochem. Biotechnol., 173*(4), 857–869.

74. Karkoula, E., Dagla, I. V., Baira, E., Kokras, N., Dalla, C., Skaltsounis, A. L., Gikas, E., & Tsarbopoulos, A., (2020). A novel UHPLC-HRMS-based metabolomics strategy enables the discovery of potential neuroactive metabolites in mice plasma, following I.P. administration of the main *Crocussativus* L. bioactive component. *J. Pharm. Biomed. Anal., 177*, 112878.

75. Chen, M., Su, M., Zhao, L., Jiang, J., Liu, P., Cheng, J., Lai, Y., et al., (2006). Metabonomic study of aristolochic acid-induced nephrotoxicity in rats. *J. Proteome Res., 5*(4), 995–1002.

76. Dong, H., Zhang, A., Sun, H., Wang, H., Lu, X., Wang, M., Ni, B., & Wang, X., (2012). Ingenuity pathways analysis of urine metabolomics phenotypes toxicity of Chuanwu in Wistar rats by UPLC-Q-TOF-HDMS coupled with pattern recognition methods. *Mol. Biosyst., 8*(4), 1206–1221.

77. Zheleva-Dimitrova, D., Simeonova, R., Gevrenova, R., Savov, Y., Balabanova, V., Nasar-Eddin, G., et al., (2019). *In vivo* toxicity assessment of *Clinopodiumvulgare* L. water extract characterized by UHPLC-HRMS. *Food Chem. Toxicol., 134*, 110841.

78. De Lima, R., Guex, C. G., Da Silva, A. R. H., Lhamas, C. L., Dos, S. M. K. L., Casoti, R., Dornelles, R. C., et al., (2018). Acute and subacute toxicity and chemical constituents of the hydroethanolic extract of *Verbenalitoralis* Kunth. *J. Ethnopharmacol., 224*, 76–84.

79. Li, X. Y., Jin, X., Li, Y. Z., Gao, D. D., Liu, R., & Liu, C. X., (2019). Network toxicology and LC-MS-based metabolomics: New approaches for mechanism of action of toxic components in traditional Chinese medicines. *Chin. Herb. Med., 11*(4), 357–363.

80. Wolfender, J. L., Terreaux, C., & Hostettmann, K., (2000). The importance of LC-MS and LC-NMR in the discovery of new lead compounds from plants. *Pharm. Biol., 38*(1), 41–54.

# CHAPTER 4

# Hyphenation of HRMS with Instruments for Phytochemical Characterization

SHINTU JUDE and SREERAJ GOPI

*Research and Development (R&D) Center, Plant Lipids (P) Ltd., Kadayiruppu, Kolenchery, Cochin, Ernakulam, Kerala – 682311, India*

## ABSTRACT

Various mass spectrometry instruments have been widely applied in research for phytochemical characterization. HRMS stands distinctly being adaptable with different kinds of ionization techniques and configurations to selectively measure the exact mass of a compound, thus offers valuable information on the physicochemical properties and characteristic minor structural changes. Besides, HRMS often hyphenated to a sensitive detector or separation technique, has provided a steady scaffold for the natural product research. In this chapter, recent advances in the applications of various separation techniques, e.g., TLC, UHPLC, SFC, CE, etc., along with detectors PDA, DAD, NMR, etc., which are hyphenated with HRMS in the context of chemical screening and identification of plant metabolites were discussed.

## 4.1 INTRODUCTION

Every analytical method or technique possess its own features and the on-line coupling of which allows a combination of analytical provisions and thereby improvement in advantages. The need for rapid and efficient strategies for screening studies has ended up in hyphenation of techniques. Hyphenation is the term denoting the conjunction of different techniques, and the establishment of hyphenated techniques opens up a possibility of customized analytical tools. In the case of High Resolution Mass

Spectrometry (HRMS), hyphenations provide extra bones in analysis strategies. HRMS provides the molecular formulae of compounds, which act as the base for compound characterization and structure elucidation. By executing fragmentations, a more detailed skeleton of the analyzing molecule is obtained. However, to complete the process and to obtain a very meaningful result, a complementary technique might require. Hyphenation doesn't place any restrictions for number of technologies to be coupled and allows the appropriate use of more than one separation or detection methods arranged in a series which allows a faster, easier, accurate identification, characterization, and full structure elucidation, and in cases, quantification also.

While dealing with the investigation of phytochemicals, the procedure involves their extraction, purification, analyzes, and structure elucidation. However, the complex matrix of plant products negatively affect the feasibility of rapid and accurate results in each and every step of processing. As we have discussed in the foregoing chapters, mass spectrometry (MS) instruments are compatible with a number of ionization sources and configurations which removes the barriers of physicochemical properties of analyte compounds in the analyzes. Thus, HRMS provides a strong platform for the natural product research in every manner, and so the hyphenation techniques.

HRMS can perform alone in a significant way in many cases. Besides, a combination of HRMS with separation techniques and even other detection techniques such as PDA, DAD, and NMR provides more promising results. Rather considering the trends and popularity, requirement plays the role here. Depending on the data needed, the hyphenations can be tailored and the possibilities are enormous.

## 4.2   BEFORE HYPHENATION: THE STAND ALONE MODE

Ambient ionization techniques are the key factor under this title. They act as the buttresses behind HRMS instruments to do the profiling analyzes by its own. Ambient ionization techniques allow different sample introduction modes which make them capable of completing the product characterization by their own. In the case of other HRMS modes, direct sample introduction is possible with infusion and it enables the direct analysis. In addition, as the name mentions, they are 'ambient'—can work under ambient conditions without much sample pretreatments. The commonly used standalone techniques are herewith discussed.

DESI-MS allows charged solvent droplets to strike on the sample surface to form the analyte ions from the surface molecules, without a pretreatment, which allows the whole sample system to remain undisturbed and undestroyed. So, it is widely used for the surface analysis, imaging, and in many cases for semi-quantitative analysis. Different herbal materials like leaf, stem, root, flowers, bark, seed, etc., were examined by DESI to characterize different types of compounds such as alkaloids, diterpenoids, diterpene glycosides, camptothecin, etc. [1–4]. On comparing the results obtained by direct ambient technique with that obtained after the extraction and LC separation were similar with reference to the number components, intensities, spectral patterns, etc. [1]. Likewise, the results from DESI detection were double confirmed for the separation and detection with TLC and TOF, respectively [3]. DESI was successfully used for demonstrating the differences in the active components content in the different organs of the same plant, considering *Nothapodytes nimmoniana* as a case [4].

While dealing with quality assurance and forensic analyzes usage, HRMS found applications in many ways. In the forensic sector, one of the major achievement induced by HRMS was, the evaluation of the cannabinoids and their derivatives. In majority of the cases, ESI-FT-ICR-MS served the purpose well. The general colorimetric screening method used for verifying the presence of cannabinoids was closely examined by ESI-FT-ICRMS/CID. By using the instrument layout, the reaction products, mechanism of reaction as well as the specificity and selectivity of the method were evaluated, and verified in presence of polyphenols from other plants also [5]. The same platform of ESI-FT-ICR MS has used for the evaluation of street samples of marijuana, which resulted in the identification of adulterants together with cannabinoids [6]. Another analytical setup of PSI-FT-ICR MS used in finding adulteration in other abused drugs such as LSD, cocaine, etc. [7]. Nonetheless, ESI-FT-ICR MS enabled the development of a platform for the prediction of plant growth time by the chemical profiling of 68 samples of cannabis seeds after cultivation, irrespective of the brand, variety, gender, or type of the seed, used for germination [8].

An important, successful maneuver furnished by DART-Orbitrap MS in the forensic division is the hair analysis for cannabinoids, which doesn't even require the sample preparation and thus allowed rapid screening [9]. DART ionization resembles the mechanism of APCI, and here, it allows the two-stage ionization on small molecular compounds from volatile samples. Moreover, the technique allows direct analyzes of samples in any form such as raw herbal materials, i.e., roots, seeds, leaves, rhizomes, fruits, etc., [10–14], herbal extracts, teas, analytes in solution, flavors, and fragrances, powdered drugs, injections, etc., and there are strategies developed according to the sample formats [15–20]. DART-HRMS was plied for the derivation of mass spectral fingerprints of the biomarker compounds from the psychoactive and medicinally important species – *Piper methysticum, Piper betle* and commercial products of the two, without any sample preparation steps [21]. The study has demonstrated that, it is possible to determine the origin of plant products by analyzing the unique chemical characteristics. Thus, even the trace amounts of adulterations in the herbal samples were illustrated by DART-MS, having cannabinoids as an example [22]. Rather, DART has successfully accomplished for drug identification [23], chemometric classification [24], reaction monitoring [17], quantification, quality control (QC) [20], etc.

DAPCI allows the ionization in a supported form from chemicals such as gasses and solvents. The scope of application ranges from component identification, quantification, QC origin differentiation, differentiation between plants [25, 26], etc. The technique can be manipulated in many ways, according to the convenience and nature of information. By altering the reagents [27] or temperature in the plasma probe [28], the results can be improved a lot. In desorption corona beam ionization (DCBI), ionization occurs by generating reactive species from the helium atoms near the corona, which in turn produce singly charged ions from sample surface. DCBI-MS techniques are used in many forensic cases to determine the adulteration in herbal medicines [29].

The standalone HRMS techniques are proved to be capable of doing the complete parameters of an analysis. However, hyphenation of MS with other techniques, especially separation techniques remarkably improves the analysis data in terms of the selectivity, efficiency, and speed. The following parts deals with some of the significant hyphenation techniques owing to HRMS as a part are depicted here with the help of a few relevant examples. For a detailed outlook, Table 4.1 is given.

**TABLE 4.1** Different Techniques Hyphenated with HRMS for Phytochemical Analysis

| Hyphenation | Benefits | References |
|---|---|---|
| TLC-HRMS | Rapid separation and identification of compounds | [32–40] |
| Frontal elution paper chromatography | Supports the direct introduction of powder sample for the elution | [41, 42] |
| UHPLC-HRMS | Qualitative and quantitative analysis within small analysis time | [46–57] |
| SFC-HRMS | Purification and analysis of chiral compounds and thermally unstable molecules | [58, 59] |
| CE-HRMS | Works well with small amount of sample, can be altered by the physicochemical properties | [63–78] |
| LC-PDA-HRMS | Effective dereplication and identification of new compounds making use of synergic effects of different detectors. | [80–88] |
| HRMS-NMR | Separation and identification along with structure elucidation of compounds | [89–109] |

## 4.3 HYPHENATION WITH SEPARATION TECHNIQUES

Natural extracts are complex matrices containing a large number of components with different natures in terms of polarity, pH, thermal sensitivity, etc. Therefore, reinforcing the HRMS readouts with a prior separation will influence the data vastly in a positive way. HRMS could be incorporated with many separation techniques such as TLC, LC, GC, SCFC, and CE. Here, in natural product analysis, GC finds a limited application due to the restrictions for samples in thermal lability and volatility. So, the other techniques are considered for the discussion.

### 4.3.1 TLC (THIN LAYER CHROMATOGRAPHY)-HRMS

TLC is a simple and rapid basic separation technique. Newer technologies were introduced to improve the efficiency of separation and resolution in TLC. The compound characterizations were made facile by coupling TLC with MS. A TLC-HRMS hyphenation is represented in Figure 4.1.

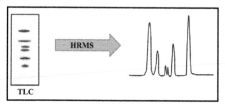

**FIGURE 4.1**    Hyphenation of TLC with HRMS.

In an old study, a hyphenation of preparative TLC with HRMS have tried, which could differentiate a number of compounds and provide data on the compound structures. It was a state of the art at that time [30]. Later, better resolutions and longer development distance were obtained by employing over pressured layer chromatography (OPLC). It is proved that, an OPLC-DART-HRMS works better than conventional HPTLC methods in terms of separation and identification of compounds [31]. However, while dealing with TLC as a separation technique in the hyphenation series, there is a need for the dilution of TLC bands to proceed further to the detections. This dilution can cause a fall in the sensitivity and low mass detection. In this regard, a different approach was introduced by Shariatgorji and his crew, where they have used a TLC-LDI (laser desorption ionization)-MS configuration, and thereby eliminated the need for any addition of another matrix [32]. In another study, the ion suppression was eliminated by introducing pre-developments of the plate, followed by HPTLC-MS$^n$ measurements. The analytes exhibited improved sensitivity as well as stability, and identified different compounds including even monomers to decamers of proanthocyanidins (PAs) [33].

Specific interaction of enzymes is used for the purification of compounds, and one such trial for the purification of β-glucosidase from *Cyamopsis tetragonoloba* was reinforced by the confirmation from HPTLC-HRMS. Here, the degree of product conversion was confirmed by HPTLC-QTOFMS and the peptide mass fingerprinting was conducted in MALDI-TOF. Thus, in a single study, two different HRMS instruments served for different purposes [34]. Even the minute differences between different species of plants belonging to the same family were distinguished by HPTLC-HRMS [35]. Similar mass fingerprinting was successfully applied to distinguish between different organs of the same plant also [36].

Interpretation of mass spectra resulting from degradation products is super clues towards compound structure. In natural product research, many strategies are developed on the same. HPTLC-HRMS data were corre-lated in such platform-eicCluster so that the otherwise hardly found mass signals of degradation products were strongly enhanced. These strength-ened signals contribute towards the structure elucidation procedures [37]. Similarly, using HPTLC-HRMS, a generic method was developed for the identification of compounds from commercially available botanical samples. Moreover, the quantification potential of the method was demon-strated with three of the active components [38]. Another hyphenation of TLC was accomplished with DESI-MS for the investigation of alkaloids from herbal dietary supplements, which could identify, characterize, and quantify the analytes, suggesting the method as a quality assurance tool [39]. Coupling of TLC to DCBI-MS was also tried in a study for the direct detection and quantification of herbal alkaloids. Here the DCBI was made strengthen by adding reactive reagent and even the low volatile species were detected rapidly [40].

In some other combinations, TLC appears as a potential part of the biological assay setup, and impart in the investigation of bioactive phyto-chemicals (refer Chapter 6). However, rather than the potentials, TLC is possessed with some disadvantages such as interfering spectral back-ground, ion suppression, need of pretreatments, etc., as discussed, some of them are solved in studies, but needed special processes or instruments for the same. These drawbacks were overwhelmed by other separation techniques.

### 4.3.2 FRONTAL ELUTION PAPER CHROMATOGRAPHY

Frontal elution paper chromatography is not much established like other separation techniques. However, in hyphenation with DCBI, it was proved to ameliorate the quality of many analyzes. It supports the powder sample to be introduced directly to the base of an isosceles triangle, which is then eluted by suitable eluent. The target analytes condenses at the tip, which is proceeded for ionization by DCBI. Figure 4.2 illustrates the working principle of frontal elution paper chromatography. The coupling allowed a rapid separation, developing, improved intensities of analytes of interest and less matrix effect. Rather, it facilitates a semi-quantitative potential, which have been demonstrated in many cases of herbal medicines and dietary supplements [41, 42].

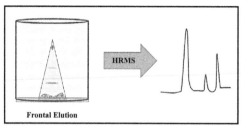

**FIGURE 4.2**    Hyphenation of frontal elution paper chromatography with HRMS.

### 4.3.3   UHPLC (ULTRAHIGH PERFORMANCE LIQUID CHROMATOGRAPHY)-HRMS

HRMS, when coupled online with the powerful and versatile separation technique UHPLC, renders miraculous results in the field of botanical investigation. It reduced the analysis time, and increased efficiency to furnish information. Figure 4.3 illustrates a simple UHPLC-HRMS hyphenation. One study took the advantage of LC-HRMS for the analysis of the neurotoxic acetogenins namely annonacin and squamocin, which were present in the lyophilized North American pawpaw (*Asimina triloba*) fruit pulp sample. They have presented the quantification of these afore-mentioned acetogenins and detection of the isomers of the same with their percentage ratio [43]. The Annonaceous acetogenin-annonacin was identified from *Annona muricata* and one of its market product samples by employing a MALDI TOF MS. It has worked as a qualitative screening tool with lesser sample preparation, analysis time and without using an internal standard [44]. The qualitative report on alkylamides was produced from *in vitro* raised plants by using LC-QTOF. The resulted correlation patterns contributed towards the format of tissue culture of bioactive sources [45].

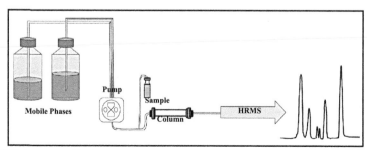

**FIGURE 4.3**    Graphical representation of UHPLC-HRMS hyphenation.

In the drift of structural identification, the most investigated samples were from the plants used for the traditional medicines. The combination of HPLC and HRMS have effectively used for the data mining of traditional Chinese medicine (TCM) formulations due to their significant influence in the present day pharma field. In such a study, three different species of the genus *Cistanche* were subjected to characterization through HPLC-LTQ-Orbitrap for herbomics research, as they are an important tonic agent in the TCM. The study successfully discovered three species from the same family, *Cistanche deserticola* Y. C. Ma, *C. tubulosa* (Schrenk) Wig and *C. sinensis* (C. A. Mey.) G. Beck to possess with a total of 69 phenylethanoid glycosides (PhGs), 17 out of them being new and 8 of them being the biomarkers [46]. Traditional medicines consist of the crude extract of single or multiple herbs, depending on the conditions. Therapeutic combinations are considered to be a more complex system. Such a complicated matrix is present in Xiao-Er-Qing-Jie (XEQJ) granules, a TCM which contains eight herbal medicines in it. Despite of the complex nature of its matrix, by using HPLC-LTQ-orbitrap XL, 91 chemical compounds were identified from the granules, including different structural moieties such as PhGs, flavonoids, phenolic acids (PA), lignans, iridoid glycosides, alkaloids, and saponins [47].

A method for the speedy, accurate detection and structural characterization of Yinchenhao Decoction (YCHD), a classical TCM formulation was developed by using HPLC-Q/TOFMS/MS, which is fit to be used as a QC aid. Find by formula (FBF) algorithms was applied for the screening of YCHD and 77 major compounds from the formulation were characterized [48]. As the first part of a combined *in vitro* and *in vivo* study on behalf of the therapeutic potential of *Bletilla striata* extracts, the ingredient structures were discerned by HPLC-ESI-HRMS [49]. A wise choice of selecting UHPLC-Q-ToF-MS as a guiding aid was made by Dorni et al. for the enrichment of triterpenoids from the ethno medically important *Centella asiatica* L. [50]. Ridder L et al. searched for the metabolic profile from green tea by applying LC-MS[n] spectral data. By using the substructure annotations, they identified 85 previously identified compounds and 24 new compounds with detailed structural information [51].

Interestingly, during the development of a method intended for an integrated identification of phytochemicals, there were many possible modes and strategies of HRMS were tried. Attempts had made to combine HRMS with GC and LC, both positive and negative modes. Using HRMS, ion source fragmentation, MS/MS fragmentation patterns, HRFS spectra, generated empirical formulae, etc., were generated and were used for the untargeted analysis [52]. Effects of extend of drying and mode of extraction on the yield, total phenolic content (TPC) and total flavonoid content (TFC) were investigated in the mastic tree leaf extracts in accordance with their antioxidant activities, and the UHPLC-HRMS identification had justified the results [53].

The administration of HRMS has been found beneficial in quantitative analyzes also. In a study, a UHPLC-Orbitrap HRMS database was developed to screen, identify, and quantify the antitussive adulterants from herbal medicines. Many possible patterns of HRMS data, such as full scan spectra, exact mass, mass fragments, isotopic data, elemental compositions, mass spectral library, etc., were incorporated in the database, in order to obtain a complete analytical identification platform for the targeted compounds. At the same time, the possibilities of HRFS and $MS^2$ HRMS data along with the assisted analysis of four software programs and a database were utilized for untargeted compounds. Subsequently, a validated quantification protocol was also established. The efficiency of the database and method as a quality assessment tool was demonstrated with 87 herbal medicine batches [54]. Application of UHPLC-HRMS coupling in quality assurance was further illustrated in finding the illegal utilization of phosphodiesterase 5 (PDE5) in natural dietary supplements. A full MS/data dependent MS/MS data acquired by Q-orbitrap was exploited for the markers identification [55]. A similar study was reported from beverages, which handled the determination of PDE5 inhibitors present in the commercial instant coffee premixes (ICPs). The process involved LC-QToFMS for the targeted screening and quantification as well as the un-targeted screening, constituting a strong platform for the adulterants determination in ICPs [56]. The same instrumental setup was utilized for the quantitative analysis of curcuminoids and their metabolites present in yet another crucial matrix-human plasma, presented with a detailed protocol and validation methodology [57].

## 4.3.4   SUPERCRITICAL FLUID CHROMATOGRAPHY (SFC)-HRMS

SFC is a modified normal phase chromatography where, the mobile phase is supercritical fluid such as $CO_2$. It finds usage in purification and analysis of chiral compounds and thermally unstable molecules with molecular mass ranging from low to medium (Figure 4.4). These properties, in combination with HRMS resulted in a versatile technique for the analysis of many natural compounds, which could otherwise be affected by the analytical conditions and parameters. One such study employed UHPSFC/QTOF-MS for the analysis of low molecular weight compounds. UHPSFC, in combination with mass instruments exhibited better performance in terms of sensitivity and matrix effects, than UHPLC [58]. Rather, it allows the ionization with different technologies. In coupling with APCI-HRMS, UHPSFC exhibited potential in screening and structure elucidation of natural, minor, and non-polar bioactive compounds [59].

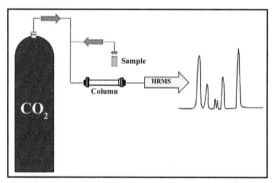

**FIGURE 4.4**   Hyphenation of SCF-HRMS.

## 4.3.5   *CAPILLARY ELECTROPHORESIS (CE)-HRMS*

As the name indicates, capillary electrophoresis (CE) is a combination capillary and electricity for separation purpose. Here, under an applied voltage, the analyte particles migrate and separate through a capillary, based on their physicochemical parameters such as charge, size, viscosity, etc. Together with HRMS, CE was found to be an intelligent combination for the analysis and characterization of compounds from natural products. A basic pattern of CE, in combination with HRMS is presented in Figure 4.5. CE can be coupled with mass spectrometers (MSs) in many formats, namely capillary gel electrophoresis (CGE), capillary zone electrophoresis (CZE), non-aqueous capillary electrophoresis (NACE), capillary isoelectric focusing (CIEF), etc., [60]. As the coupling of CE with ESI is more common, the potential of this hyphenation was tailored by including an interface, which can be assigned as sheathless, sheath-liquid, or liquid junction interfaces. Sheathless interface couples CE capillary to the MS directly, maintaining the electrical contact by a conductive metal, an electrode or by a spraying tip. The performance of the system in this case, can be improved by playing with the ESI emitter design, selection of the proper buffer solution, etc. In sheath-liquid interfaces, the separated liquid from CE is combined coaxially with the sheath flow liquid flowing through a capillary or transfer line directing towards the source. The presence of a makeup liquid differentiates liquid-liquid interface [61].

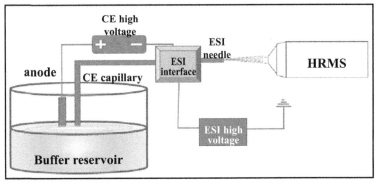

FIGURE 4.5   Schematic diagram of CE-HRMS hyphenation.

Requirement of very small amount of sample for the analyzes in CE-HRMS combination made it a suitable platform for many kind of analysis patterns such as process development, non-targeted profiling, metabolomics, proteomics, qualitative, and quantitative. Non-targeted profiling and the quantitative data of complete metabolites have a significant role in new product researches also. For example, CE-DAD-HRMS was successfully applied for the identification of alkaloids from *R. coptidis* and quantitative determination of the major three among them. In this study, the compound identification was carried out with both CE-TOF-MS with HPLC-TOF-MS, and demonstrated that, on combining with different separation techniques, these two couplings act as complementary to each other [62]. Cyclodextrin based CZE was developed for the determination of intact glucosinolates (GSL), and was employed for the analysis of broccoli based dietary supplements. Considering GSL as the reference, the analysis layout provided a platform for quality assurance [63]. In another case, the profiling of *Stemona* alkaloids was accomplished along with their quantification and fragmentation patterns, by using a NACE-ESI-IT-MS having a sheath liquid interface [64]. A proper differentiation between different isoforms formed by the glycosylation of cellobiohydrolase 1 was achieved by the coupling of gel isoelectric focusing and CE with HRMS [65].

Both qualitative and quantitative profiling of a number of active constituents from herbal extracts was fulfilled by CE-HRMS. By using CE-TOF-MS, the changes of sensory attributes and quality parameters, which are specific to the storage conditions of vegetable bean edamame were proved to be related to its change in metabolic profiles, especially in the amino acid levels [66]. The metabolic study of pineapple leaves demonstrated Crassulacean acid metabolism by the simultaneous evaluation of carboxylic acid and amino acids [67]. In another study, CE-TOF-MS in combination with principal component analysis (PCA) was used for the identification of metabolite profiling and characterization of the six different herbs included in a TCM named *Toki-Shakuyaku-San* [68]. Different kinds of fruits and fruit products were examined with the CE-HRMS system and the data found useful to identify the fruit of origin, which has led to an important finding of adulteration in a fruit product from the market [69]. A number of varieties of flavan-3-ols were identified in the seeds from the pomace of the red grape vinification of *Vitis vinifera* (Cabernet Sauvignon) by using chiral CE and LC-ESI-FTICR-MS. Among the phenolic compounds in grape seeds, 251 different flavan-3-ol compounds were distinctively identified with elemental composition and the privilege of chiral CE resulted in identifying even the exact enantiomers resulted from wine making [70]. Besides, the same instrument combination was successfully utilized for the qualitative and quantitative determination of isoquinoline alkaloids present in *Corydalis* species. PCA processing of the data allowed the production lot discrimination of samples [71]. The CE-HRMS supported analysis of *Arabidopsis* extracts provided favorable conditions for minimized ion suppression and isomer separation, along with a better limit of quantitation like 80 nM of the analyte from 33 pmol/g of fresh plant weight [72].

By including an enzymatic digestion to the CE-UHPLC-HRMS hyphenation, it is possible to analyze even large proteins [73]. By immobilizing the enzymes on the column capillary wall, protein diffusion is achieved within a short span of time, with very low flow rate. The mode of immobilization can be differed such as specific or non-specific adsorption, magnetic, non-magnetic, monolithic, or covalent packing, narrow bore capillaries or microfluidic channels, etc., [74–77]. In connection with HRMS, they are proved to be efficient enough for the loading of small amount of proteins and the online coupling is associated with lesser time of analysis [77]. Furthermore, the separation could achieve up to proteoform level, allowing the accurate assignment of protein isoforms [78].

## 4.4   HRMS IN HYPHENATION WITH OTHER DETECTORS

The conjunction of more than one detector would allow a synergic effect in the HPLC results, as the detectors complement each other. One such attempt was the phenolic profiling study on the matrices involved in the oil processing of *Olea europaea* L. (European olive) using UHPLC-DAD-ESI-QTOF combination. Six matrices were subjected to the analysis using HRMS-QTOF Synapt MS, from which, 80 different polyphenols were identified and characterized on the basis of analysis data [79]. A research on *Brassica napus* L. var napus (rapeseed) tried to identify the phenolic compounds in the crude methanol extract of the seeds by using HPLC-PDA-ESI (-)-$MS^n$/HRMS. 91 flavonoids and hydroxyl cinnamic acid derivatives were detected in the analysis, proving that, HPLC-PDA–ESI (–)-$MS^n$/HRMS (QTOF) combination is a highly efficient technique for chemical identification and plant phenolic analysis [80]. Lin et al. identified and characterized 209 different phenolic compounds from *Brassica juncea* Coss variety (red mustard green) using a platform of UHPLC-PDA-ESI/HRMS/ $MS^n$. They have used an LTQ Orbitrap XL MS for the successful screening of compounds including anthocyanins, flavonol glycosides, and hydroxycinnamic acid derivatives out of which almost 100 compounds were reporting for the first time in Brassica plants [81]. A series of analysis methods were derived for the bioactive components of the medicinal mushroom *Antrodia cinnamomea*, making use of UHPLC/DAD/qTOF-MS, UHPLC/UV, supercritical fluid chromatography (SFC)-MS, and ion chromatography coupled with pulsed amperometric detection (IC/PAD). A QC device for the vital triterpenoids ergostane and lanostane was prepared and applied for 15 batches of the species [82].

Characterization of cytotoxic secondary metabolites from the crude extracts of small scale fungal cultures were proved to be possible by a UHPLC-PDA-HRMS-MS/MS methodology. A short timed qualitative analysis has developed with high resolution and mass accuracy using LTQ Orbitrap XL MS which results in the detection and identification of trace amount of compounds in the crude extracts [83]. Another work reported the antifungal activity of the roasted hazelnut (*Corylus avellana* L.) skin (RHS) extract and its sub-fractions against *Candida albicans* SC5314 pathogenicity. The polyphenols, mainly the bioactive PAs involved in the activity are chemically characterized using a combination of HPLC-UV and HRMS. The analysis setup consisting of linear ion trap (IT)-Orbitrap hybrid mass spectrometer (LTQ OrbiTrap XL) for both the direct flow injection analysis (FIA), and chromatographic analysis was appropriately used for portraying the metabolic profile and structure elucidation of the determined PA types [84]. An endophytic fungal named *Aspergillus iizukae*, isolated from the leaves of milk thistle (*Silybum marianum* L.), was proved to contain three flavonolignans-Silybin A, silybin B, and isosilybin A, with the help of LTQ Orbitrap XL mass spectrometer, hyphenated in UHPLC-PDA-HRMS-MS/MS. The isolated compounds were found to be the same compounds of their host plants and this vital information could pave way for many chemical and evolutionary studies [85].

A droplet-liquid microjunction-surface sampling probe (droplet-LMJ-SSP) was coupled with UHPLC-PDA-HRMS-MS/MS, in order to develop a protocol for the analysis of secondary metabolites from fungal cultures. A set of mutually supportive data, such as separation, RT, MS data, and UV/vis data were collected by exploring the analysis setup including DESI-MS. The setup enabled the dereplication of different fungal cultures along with their identification, separation of isomers and mapping of secondary metabolites without any sample preparation [86].

Extraction conditions are sometimes the most important part of process optimization. With the help of TLC-HPLC-ELSD-DAD-HRMS, a simple optimization procedure was demonstrated for the pressurized fluid extraction (PFE) of *Eugenia uniflora* L. The chemical composition and biochemical activity were examined with the same instrumental setup [87]. A modified version of the above-mentioned hyphenation was fabricated as HPTLC-HPLC-DAD-ELSD-UHPLC-HRMS-GC-MS, for the fingerprinting of four plant organs of hybrid rose variety 'Jardin de Granville,' consisting of a wide range of molecular families. A subsidiary data of polarity, existing plant organ, and extraction solvent corresponding to each compound is provided along with a platform of quality assurance parameters [88].

## 4.5  HRMS-NMR AND MORE

HRMS and its hyphenation with other techniques were evidenced as potent contributors in the structural characterization of compounds, even if they are unknown to date or present in low concentrations in the sample. Coupling of HRMS with NMR, the universal detector was a turning point in the field of structural analysis. The range of structural identification provided by the combo was beyond expectation-complete structural characterization for the isolated compound was made possible.

Some studies are found to be interested as they do not search for just the components present in the plants, but give an idea about the activities of compounds present in it. Some of such studies are tried to discuss in this section. *Pseudoxylaria sp.* X802, generally considered as a stowaway fungus was investigated for their capability to produce bioactive metabolites during the co-cultivation with different fungi. Using the NMR-MALDI-TOFMS analysis identified the structural characteristics of six new cyclotetrapeptides, pseudoxylallemycins A–F (1–6) [89]. A novel limonoid named Tooniliatone A, with an unprecedented 6/5/6/5 tetracarbocyclic framework was isolated from *Toonaciliata Roem.* var. yunnanensis. LC-HRMS-NMR was used to characterize the structure and to confirm it as a genuine natural product [90]. In another study, three new sesquiterpene lactone dimers named dicarabrol A, dicarabrone C and dipulchellin A were identified from the whole plants of *Carpesium abrotanoides*. The structure elucidation was completed with the help of NMR, HRESIMS, and X-ray crystallography [91]. Sixteen new limonoids named Entangolensins A-P, having great variety in the frameworks were distinguished from the stem barks of *Entandrophragma angolense* and the structure were elucidated by HPLC-HRMS-NMR-ECD [92].

A research has proved the aforementioned fact by investigating a crude ethanol extract of *Carthamus oxyacantha* M (wild safflower) for the screening of compounds using two methodologies-traditional fractionations by VLC followed by HPLC and another series consisting HRMS as HPLC-PDA-HRMS-SPE-NMR. By including SPE (solid phase extraction) fractionation in the workflow, the whole process was accelerated by the reduction of interfering compounds. Moreover, the system was successful in the identification of 15 compounds and was beneficial by means of labor, cost, and time [93]. In another case, the identification and structural characterization of isobaric iridoid glycoside regioisomers from *Harpagophytum procumbens* DC was conducted by LC-DAD-MS/SPE-NMR [94].

In a recent study, nineteen known compounds along with two new isoflavanones were isolated from *Erythrina brucei*. The structural characterizations were made with NMR-CD-IR-ESI-HRMS$^n$ configuration [95]. Han et al. presents the isolation of two novel alkaloids-robustanoids A and B from *Hainan robusta* coffee (*Coffea canephora*) beans and the complete structure elucidation using HRESIMS-IR-NMR-electronic circular dichroism (ECD). The structure was confirmed by a total synthesis of both the compounds. They have also successfully proved the comparable α-glucosidase inhibitory activity of robustanoids B [96]. Comprehensive chemical characteristic information on *Molopanthera paniculata* Turcz, gathered with HRMS-NMR could support the recent classification of *Posoqueria* Aubl. and *Molopanthera* in a new single tribe *Posoquerieae* [97].

A double HRMS study was conducted for the structural identification of eleven newly isolated phthalide derivatives from the rhizome of *Ligusticum chuanxiong*. All the structures were ascertained by UV, IR, HPLC-Q-TOF, LCMS-IT-TOF, NMR, and ECD spectra [98]. NMR-HRESIMS-UV-ECD-IR hyphenated analysis setup was used for the structure elucidation of nine new compounds discovered from the roots of *Lycium chinense* Mill. Determined the α-glucosidase inhibitory activity of new compounds including 1 flavane, 1 amide, 1 sesquiterpene, 3 lignin glucosides, and three phenolic glucosides [99]. A bioactivity guided isolation of oxypregnane-oligoglycosides, commonly known as calotroposides was carried out from the roots of *Calotropis gigantea* (L.), in search of anti-cancer drugs. NMR-HRMS based dereplication of the study material provided the dataset of one new and six known calotroposides, and the structure elucidation was completed by HPLC-UV-IR-HRMS-NMR [100]. Two rare *Chloranthus* species-*C. oldhamii* Solms-Laub and *C. sessilifolius* K. F. Wu – have been investigated for their signature compounds and the isolated marker compounds found to have a distinctive structural framework was studied for the anti-neuroinflammatory activity [101].

The chemistry of new compounds identification has advanced to a greater distance by the introduction of HRMS-NMR combination. One example was the identification of new bioactive secondary metabolites from *Platanus* species. The isolation and structure elucidation were carried out by the instrument combination [102]. In the same way, 17 new compounds were isolated from *Citharexylum spinosum* L., four of them being remained undescribed to that time. Yet, their isolation and structural characterization were carried out by UHPLC-HRMS-NMR [103]. Following the similar pattern, alkaloids from *Narcissus pseudonarcissus* L. cv. Dutch Master fresh bulbs, triterpenes from *Echinops spinosissimus* Turra subsp. *spinosissimus*, phenolic compounds from *Rhodiola imbricate*, isoflavans from *Erythrina livingstoniana*, flavanones from *Erythrina livingstoniana*, etc., were characterized [104–108]. In case of *Erythrina livingstoniana*, along with structure elucidation, different possible biosynthetic pathways were also established [107]. Differentiation of the compounds from different organ tissues of the same plant is another notable achievement by using UHPLC-HRMS NMR [109].

A progressive and complementary screening strategy was introduced by combining HPTLC-HPLC-DAD-HPLC-ESI-HRMS for the phytochemical characterization. The rapid access to the molecular classes by HPTLC guided HPLC analysis of polyphenol compounds in a highly resolved and refined way. DAD recorded characteristic absorbance of each corresponding compound, while HRMS allowed the mass detection and identification of compound [110]. The same instrumentation was used in a different format in another study where the TLC and HPLC-DAD parts served for the fractionation and semi-preparative isolation of compounds. Further, HPLC-PDA-SPE-NMR in two configurations was performed, in the first case, it was a purification technology, and in the second case, it has coupled with HRMS and proceeded for mass measurements towards compound identification. In each step, NMR played a role in structure elucidation [111]. Both identification and structure elucidation of major constituents from *Spondias tuberosa* fruits was accomplished by different configurations of HPLC-PDA-ESI-MS and UHPLC-TOF-HRMS along with NMR [112].

Identification and structure elucidation of compounds provide many details on the nature and activities of the same. In many studies, this kind of correlation was made, rendering a wide view on the properties. The antioxidant activities and polyphenolic content of *Castanea sativa* were paralleled with its phytochemical contents as profiled by LC-ESI/

LTQOrbitrap/MS/MS$^{n}$-NMR. Isolation, class identification and structure elucidation-everything have executed by the single hyphenation [113]. Another important identification process made by LC-HRMS/MS-NMR is in enzymatic hydrolysis of oleuropein. The regioselective hydrolysis using two different enzyme preparations had produced different biological active compounds according to the substrate specificity of enzymes and these tailoring were confirmed by the instrument hyphenation [114]. Similarly, the structure elucidation, along with the determination of absolute configuration was another important application. Differentiation of diastereomers is substantial, as the defined activities will be different for them and hence the therapeutic decisions can be evaluated precisely without errors [115]. During the research on *Salaciastaudtiana* Loes. ex Fritsch., many active compounds were isolated and identified, and their structures were elucidated with relative configurations by HRMS-NMR-XRD. The structure-activity relationship (SAR) of the active compounds was also established [116]. Configuration assignment could done up to epimer differentiation [117].

An HRMS-based metabolic platform and biochemometric statistical approach were bridged with UHPLC-UV-HRMS, in order to identify the possible anti-cholinesterase compounds from different *Zanthoxylum* species. Findings from this approach were further verified by the bio-directed isolation [118].

## 4.6  CONCLUSION

Hyphenation of instruments provides a multifold improvement in producing data, due to the synergic effect of techniques used. Most of the acclaimed properties of HRMS, such as sensitivity, resolution, isomers separation, etc., are enhanced by coupling with other separation techniques. In combination with other detectors, HRMS allows the complete characterization of the analytes of interest, including their structure elucidation, even up to confirmations. Altogether, the hyphenation of instruments keeps them in an advanced position in the natural product research and drug discovery programs.

## KEYWORDS

- capillary electrophoresis
- capillary isoelectric focusing
- desorption corona beam ionization
- electronic circular dichroism
- frontal elution paper chromatography
- hyphenation

## REFERENCES

1. Talaty, N., Takáts, Z., & Cooks, R. G., (2005). Rapid in situ detection of alkaloids in plant tissue under ambient conditions using desorption electrospray ionization. *Analyst, 130*(12), 1624–1633.

2. Jackson, A. U., Tata, A., Wu, C., Perry, R. H., Haas, G., West, L., & Cooks, R. G., (2009). Direct analysis of Stevia leaves for diterpene glycosides by desorption electrospray ionization mass spectrometry. *Analyst, 134*(5), 867–874.

3. Kennedy, J. H., & Wiseman, J. M., (2010). Direct analysis of *Salvia divinorum* leaves for salvinorin A by thin layer chromatography and desorption electrospray ionization multi-stage tandem mass spectrometry. *Rapid Commun. Mass Spectrom., 24*(9), 1305–1311.

4. GrahamáCooks, R., (2011). Direct analysis of camptothecin from nothapodytes nimmoniana by desorption electrospray ionization mass spectrometry (DESI-MS). *Analyst, 136*(15), 3066–3068.

5. Dos, S. N. A., Souza, L. M., Domingos, E., França, H. S., Lacerda, Jr. V., Beatriz, A., Vaz, B. G., et al., (2016). Evaluating the selectivity of colorimetric test (Fast Blue BB salt) for the cannabinoids identification in marijuana street samples by UV-Vis, TLC, ESI (+) FT-ICR MS and ESI (+) MS/MS. *Forensic Chem., 1*, 13–21.

6. Nascimento, I. R., Costa, H. B., Souza, L. M., Soprani, L. C., Merlo, B. B., & Romão, W., (2015). Chemical identification of cannabinoids in street marijuana samples using electrospray ionization FT-ICR mass spectrometry. *Food Anal. Methods, 7*(4), 1415–1424.

7. Allochio, F. J. F., Lacerda, Jr. V., & Romão, W., (2019). Fourier transform mass spectrometry applied to forensic chemistry. In: Schmitt-Kopplin, P., & Kanawati, B., (eds.), *Fundamentals and Applications of Fourier Transform Mass Spectrometry* (pp. 469–508). Elsevier: USA.

8. Borille, B. T., Ortiz, R. S., Mariotti, K. C., Vanini, G., Tose, L. V., Filgueiras, P. R., Marcelo, M. C., et al., (2017). Chemical profiling and classification of cannabis through electrospray ionization coupled to Fourier transform ion cyclotron resonance mass spectrometry and chemometrics. *Food Anal. Methods, 9*(27), 4070–4081.

9. Duvivier, W. F., Van, B. T. A., Pennings, E. J., & Nielen, M. W., (2014). Rapid analysis of Δ-9-tetrahydrocannabinol in hair using direct analysis in real time ambient ionization orbitrap mass spectrometry. *Rapid Commun. Mass Spectrom., 28*(7), 682–690.

10. Banerjee, S., Madhusudanan, K. P., Chattopadhyay, S. K., Rahman, L. U., & Khanuja, S. P., (2008). Expression of tropane alkaloids in the hairy root culture of *Atropa acuminata* substantiated by DART mass spectrometric technique. *Biomed. Chromatogr., 22*(8), 830–834.

11. Lesiak, A. D., Cody, R. B., Dane, A. J., & Musah, R. A., (2015). Plant seed species identification from chemical fingerprints: A high-throughput application of direct analysis in real time mass spectrometry. *Anal. Chem., 87*(17), 8748–8757.

12. Kumar, S., Bajpai, V., Singh, A., Bindu, S., Srivastava, M., Rameshkumar, K. B., & Kumar, B., (2015). Rapid fingerprinting of Rauwolfia species using direct analysis in real time mass spectrometry combined with principal component analysis for their discrimination. *Food Anal. Methods., 7*(14), 6021–6026.

13. Kim, H. J., & Jang, Y. P., (2009). Direct analysis of curcumin in turmeric by DART-MS. *Phytochem. Anal., 20*(5), 372–377.

14. Kim, H. J., Baek, W. S., & Jang, Y. P., (2011). Identification of ambiguous cubeb fruit by DART-MS-based fingerprinting combined with principal component analysis. *Food Chem., 129*(3), 1305–1310.

15. Wang, Y., Li, C., Huang, L., Liu, L., Guo, Y., Ma, L., & Liu, S., (2014). Rapid identification of traditional Chinese herbal medicine by direct analysis in real time (DART) mass spectrometry. *Analytica Chimica Acta, 845*, 70–76.

16. Shen, Y., Van, B. T. A., Claassen, F. W., Zuilhof, H., Chen, B., & Nielen, M. W., (2012). Rapid control of Chinese star anise fruits and teas for neurotoxic anisatin by direct analysis in real-time high-resolution mass spectrometry. *J. Chromatogr. A, 1259*, 179–186.

17. Petucci, C., Diffendal, J., Kaufman, D., Mekonnen, B., Terefenko, G., & Musselman, B., (2007). Direct analysis in real time for reaction monitoring in drug discovery. *Anal. Chem., 79*(13), 5064–5070.

18. Haefliger, O. P., & Jeckelmann, N., (2007). Direct mass spectrometric analysis of flavors and fragrances in real applications using DART. *Rapid Commun. Mass Spectrom., 21*(8), 1361–1366.

19. Steiner, R. R., & Larson, R. L., (2009). Validation of the direct analysis in real time source for use in forensic drug screening. *J. Forensic Sci., 54*(3), 617–622.

20. Li, Y. J., Wang, Z. Z., Bi, Y. A., Ding, G., Sheng, L. S., Qin, J. P., Xiao, W., et al., (2012). The evaluation and implementation of direct analysis in real time quadrupole time-of-flight tandem mass spectrometry for characterization and quantification of geniposide in Re Du Ning Injections. *Rapid Commun. Mass Spectrom., 26*(11), 1377–1384.

21. Bajpai, V., Sharma,D., Kumar, B., & Madhusudanan, K. P., (2010). Profiling of *Piper betle* Linn. Cultivars by direct analysis in real time mass spectrometric technique. *Biomed. Chromatogr., 24*(12), 1283–1286.

22. Musah, R. A., Domin, M. A., Walling, M. A., & Shepard, J. R., (2012). Rapid identification of synthetic cannabinoids in herbal samples via direct analysis in real time mass spectrometry. *Rapid Commun. Mass Spectrom., 26*(9), 1109–1114.

23. Samms, W. C., Jiang, Y. J., Dixon, M. D., Houck, S. S., & Mozayani, A., (2011). Analysis of alprazolam by DART-TOF mass spectrometry in counterfeit and routine drug identification cases. *J. Forensic Sci., 56*(4), 993–998.

24. Lee, S. M., Kim, H. J., & Jang, Y. P., (2012). Chemometric classification of morphologically similar umbelliferae medicinal herbs by DART-TOF-MS fingerprint. *Phytochem. Anal., 23*(5), 508–512.

25. Zhang, X., Jia, B., Huang, K., Hu, B., Chen, R., & Chen, H., (2010). Tracing origins of complex pharmaceutical preparations using surface desorption atmospheric pressure chemical ionization mass spectrometry. *Anal. Chem., 82*(19), 8060–8070.

26. Pi, Z., Yue, H., Ma, L., Ding, L., Liu, Z., & Liu, S., (2011). Differentiation of various kinds of *Fructus schisandrae* by surface desorption atmospheric pressure chemical ionization mass spectrometry combined with principal component analysis. *Analytica Chimica Acta, 706*(2), 285–290.

27. Song, Y., & Cooks, R. G., (2006). Atmospheric pressure ion/molecule reactions for the selective detection of nitroaromatic explosives using acetonitrile and air as reagents. *Rapid Commun. Mass Spectrom., 20*(20), 3130–3138.

28. Liu, Y., Lin, Z., Zhang, S., Yang, C., & Zhang, X., (2009). Rapid screening of active ingredients in drugs by mass spectrometry with low-temperature plasma probe. *Anal. Bioanal. Chem., 395*(3), 591–599.

29. Yang, Y., & Deng, J., (2016). Analysis of pharmaceutical products and herbal medicines using ambient mass spectrometry. *Trends Anal. Chem., 82*, 68–88.

30. Seifert, W. K., & Teeter, R. M., (1969). Preparative thin-layer chromatography and high-resolution mass spectrometry of crude oil carboxylic acids. *Anal. Chem., 41*(6), 786–795.

31. Móricz, Á. M., Häbe, T. T., Ott, P. G., & Morlock, G. E., (2019). Comparison of high-performance thin-layer with over pressured layer chromatography combined with direct bioautography and direct analysis in real time mass spectrometry for tansy root. *J. Chromatogr. A, 1603*, 355–360.

32. Shariatgorji, M., Spacil, Z., Maddalo, G., Cardenas, L. B., & Ilag, L. L., (2009). Matrix-free thin-layer chromatography/laser desorption ionization mass spectrometry for facile separation and identification of medicinal alkaloids. *Rapid Commun Mass Spectrom., 23*(23), 3655–3660.

33. Glavnik, V., Vovk, I., & Albreht, A., (2017). High performance thin-layer chromatography-mass spectrometry of Japanese knotweed flavan-3-ols and proanthocyanidins on silica gel plates. *J. Chromatogr A, 1482*, 97–108.

34. Asati, V., & Sharma, P. K., (2019). Purification and characterization of an isoflavones conjugate hydrolyzing β-glucosidase (ICHG) from *Cyamopsis tetragonoloba* (guar). *Biochem. Biophys. Rep., 20*, 100669.

35. Hage, S., & Morlock, G. E., (2017). Bioprofiling of Salicaceae bud extracts through high-performance thin-layer chromatography hyphenated to biochemical, microbiological and chemical detections. *J. Chromatogr. A, 1490*, 201–211.

36. Orsini, F., Vovk, I., Glavnik, V., Jug, U., & Corradini, D., (2019). HPTLC, HPTLC-MS/MS and HPTLC-DPPH methods for analyses of flavonoids and their antioxidant activity in *Cyclanthera pedata* leaves, fruits and dietary supplement. *J. Liq. Chromatogr. Relat. Technol.42*(9/10), 290–301.

37. Fichou, D., Yüce, I., & Morlock, G. E., (2018). eicCluster software, an open-source *in silico* tool, and on-surface syntheses, an *in-situ* concept, both exploited for signal highlighting in high-resolution mass spectrometry to ease structure elucidation in planar chromatography. *J. Chromatogr. A, 1577*, 101–108.

38. Krüger, S., Hüsken, L., Fornasari, R., Scainelli, I., & Morlock, G. E., (2017). Effect-directed fingerprints of 77 botanicals via a generic high-performance thin-layer chromatography method combined with assays and mass spectrometry. *J. Chromatogr., 1529*, 93–106.

39. Van, B. G. J., Tomkins, B. A., & Kertesz, V., (2007). Thin-layer chromatography/desorption electrospray ionization mass spectrometry: Investigation of goldenseal alkaloids. *Anal. Chem., 79*(7), 2778–2789.

40. Hou, Y., Wu, T., Liu, Y., Wang, H., Chen, Y., Chen, B., & Sun, W., (2014). Direct analysis of quaternary alkaloids by in situ reactive desorption corona beam ionization MS. *Analyst., 139*(20), 5185–5191.

41. Huang, Y. Q., You, J. Q., Zhang, J., Sun, W., Ding, L., & Feng, Y. Q., (2011). Coupling frontal elution paper chromatography with desorption corona beam ionization mass spectrometry for rapid analysis of chlorphenamine in herbal medicines and dietary supplements. *J. Chromatogr. A, 1218*(41), 7371–7376.

42. Huang, Y. Q., You, J. Q., Cheng, Y., Sun, W., Ding, L., & Feng, Y. Q., (2013). Frontal elution paper chromatography for ambient ionization mass spectrometry: Analyzing powder samples. *Food Anal. Methods., 5*(16), 4105–4111.

43. Levine, R. A.,Richards, K. M., Tran, K., Luo, R., Thomas, A. L., & Smith, R. E., (2015). Determination of neurotoxic acetogenins in pawpaw (*Asimina triloba*) Fruit by LC-HRMS. *J. Agric. Food Chem., 63*, 1053–1056.

44. Champy, P., Guérineau, V., & Laprévote, O., (2009). MALDI-TOF MS profiling of annonaceous acetogenins in *Annona muricata* products for human consumption. *Molecules, 14*(12), 5235–5246.

45. Bhat, Z. S., Jaladi, N., Khajuria, R. K., Shah, Z. H., & Arumugam, N., (2016). Comparative analysis of bioactive N-alkylamides produced by tissue culture raised versus field plantlets of *Spilanthes ciliata* using LC-Q-TOF (HRMS). *J. Chromatogr. B., 1017*, 195–203.

46. Zhang, J., Li, C., Che, Y., Wu, J., Wang, Z., Cai, W., Li, Y., M, Z., & Tu, P., (2015). LTQ-Orbitrap-based strategy for traditional Chinese medicine targeted class discovery, identification, and herbomics research: A case study on phenylethanoid glycosides in three different species of herbs *Cistanches. RSC Adv., 5*, 80816–80828.

47. Li, Y., Liu, Y., Liu, R., Liu, S., Zhang, X., Wang, Z., Zhang, J., & Lu, J., (2015). HPLC-LTQ-orbitrap MSn profiling method to comprehensively characterize multiple chemical constituents in Xiao-Er-Qing-Jie granules. *Anal. Methods, 7*, 7511–7526.

48. Fu, Z., Li, Z., Hu, P., Feng, Q., Xue, R., Hu, Y., & Huang, C., (2015). A practical method for the rapid detection and structural characterization of major constituents from traditional Chinese medical formulas: Analysis of multiple constituents in yinchenhao decoction. *Anal. Methods, 7*, 4678–4690.

49. Wang, Y., Huang, W., Zhang, J., Yang, M., Qi, Q., Wang, K., Li, A., & Zhao, Z., (2016). The therapeutic effect of *Bletilla striata* extracts on LPS-induced acute lung injury by the regulations of inflammation and oxidation. *RSC Adv., 6*(92), 89338–89346.

50. Dorni, A. I. C., Peter, G., Jude, S., Arundhathy, C. A., Jacob, J., Amalraj, A., Pius, A., & Gopi, S., (2017). UHPLC–Q-ToF-MS-guided enrichment and purification of triterpenoids from *Centella asiatica* (L.) extract with macroporous resin. *J. Liq. Chrom. Rel. Technol., 40*(1) 13–25.

51. Ridder, L., Van, D. H. J. J., Verhoeven, S., De Vos, R. C., Bino, R. J., & Vervoort, J., (2013). Automatic chemical structure annotation of an LC-MS$^n$ based metabolic profile from green tea. *Anal Chem., 18*(85), 6033–6040.

52. Ballesteros-Vivas, D., Álvarez-Rivera, G., Sánchez-Camargo, A. D. P., Ibáñez, E., Parada-Alfonso, F., & Cifuentes, A., (2019). A multi-analytical platform based on pressurized-liquid extraction, *in vitro* assays and liquid chromatography/gas chromatography coupled to high resolution mass spectrometry for food by-products valorization. Part 1: Withanolides-rich fractions from golden berry (*Physalis peruviana* L.) calyces obtained after extraction optimization as case study. *J. Chromatogr. A., 1584*, 155–164.

53. Bampouli, A., Kyriakopoulou, K., Papaefstathiou, G., Louli, V., Aligiannis, N., Magoulas, K., & Krokida, M., (2015). Evaluation of total antioxidant potential of *Pistacia lentiscus* Var. chia leaves extracts using UHPLC–HRMS. *J. Food Eng., 167*, 25–31.

54. Guo, C., Gong, L., Wang, W., Leng, J., Zhou, L., Xing, S., Zhao, Y., Xian, R., Zhang, X., & Shi, F., (2020). Rapid screening and identification of targeted or non-targeted antitussive adulterants in herbal medicines by Q-Orbitrap HRMS and screening database. *Int. J. Mass Spectrom., 447*, 116250.

55. Jiru, M., Stranska-Zachariasova, M., Dzuman, Z., Hurkova, K., Tomaniova, M., Stepan, R., Cuhra, P., & Hajslova, J., (2019). Analysis of phosphodiesterase type 5 inhibitors as possible adulterants of botanical-based dietary supplements: Extensive survey of preparations available at the Czech market. *J. Pharm. Biomed. Anal., 164*, 713–724.

56. Yusop, A. Y. M., Xiao, L., & Fu, S., (2019). Determination of phosphodiesterase 5 (PDE5) inhibitors in instant coffee premixes using liquid chromatography-high-resolution mass spectrometry (LC-HRMS). *Talanta, 204*, 36–43.

57. Jude, S., Amalraj, A., Kunnumakkara, A. B., Divya, C., Löffler, B. M., & Gopi, S., (2018). Development of validated methods and quantification of curcuminoids and curcumin metabolites and their pharmacokinetic study of oral administration of complete natural turmeric formulation (Cureit™) in human plasma via UPLC/ESI-Q-TOF-MS spectrometry. *Molecules, 23*(10), 2415.

58. Storbeck, K. H., Gilligan, L., Jenkinson, C., Baranowski, E. S., Quanson, J. L., Arlt, W., & Taylor, A. E., (2018). The utility of ultra-high performance supercritical fluid chromatography-tandem mass spectrometry (UHPSFC-MS/MS) for clinically relevant steroid analysis. *J. Chromatogr. B., 1085*, 36–41.

59. Duval, J., Colas, C., Pecher, V., Poujol, M., Tranchant, J. F., & Lesellier, E., (2017). Hyphenation of ultra-high performance supercritical fluid chromatography with atmospheric pressure chemical ionization high-resolution mass spectrometry: Part 1. Study of the coupling parameters for the analysis of natural non-polar compounds. *J. Chromatogr. A., 1509*, 132–140.

60. Stolz, A., Jooß, K., Höcker, O., Römer, J., Schlecht, J., & Neusüß, C., (2019). Recent advances in capillary electrophoresis-mass spectrometry: Instrumentation, methodology and applications. *Electrophoresis, 40*(1), 79–112.

61. Tomer, K. B., (2001). Separations combined with mass spectrometry. *Chem. Rev., 101*(2), 297–328.

62. Chen, J., Zhao, H., Wang, X., Lee, F. S. C., Yang, H., & Zheng, L., (2008). Analysis of major alkaloids in rhizoma coptidis by capillary electrophoresis-electrospray-time of flight mass spectrometry with different background electrolytes. *Electrophoresis, 29*(10), 2135–2147.

63. Lechtenberg, M., & Hensel, A., (2019). Determination of glucosinolates in broccoli-based dietary supplements by cyclodextrin-mediated capillary zone electrophoresis. *J. Food Compos. Anal., 78*, 138–149.

64. Sturm, S., Schinnerl, J., Greger, H., & Stuppner, H., (2008). Nonaqueous capillary electrophoresis-electrospray ionization-ion trap-mass spectrometry analysis of pyrrolo-and pyrido [1, 2-a] azepine alkaloids in *Stemona. Electrophoresis, 29*(10), 2079–2087.

65. Sandra, K., Stals, I., Sandra, P., Claeyssens, M., Van, B. J., & Devreese, B., (2004). Combining gel and capillary electrophoresis, nano-LC and mass spectrometry for the elucidation of post-translational modifications of *Trichodermareesei* cellobiohydrolase I. *J. Chromatogr. A., 1058*(1/2), 263–272.

66. Sugimoto, M., Goto, H., Otomo, K., Ito, M., Onuma, H., Suzuki, A., Sugawara, M., et al., (2010). Metabolomic profiles and sensory attributes of edamame under various storage duration and temperature conditions. *J. Agric. Food Chem., 58*(14), 8418–8425.

67. Wakayama, M., Aoki, N., Sasaki, H., & Ohsugi, R., (2010). Simultaneous analysis of amino acids and carboxylic acids by capillary electrophoresis-mass spectrometry using an acidic electrolyte and uncoated fused-silica capillary. *Anal. Chem., 82*(24), 9967–9976.

68. Iino, K., Sugimoto, M., Soga, T., & Tomita, M., (2012). Profiling of the charged metabolites of traditional herbal medicines using capillary electrophoresis time-of-flight mass spectrometry. *Metabolomics., 8*(1), 99–108.

69. Navarro, M., Núñez, O., Saurina, J., Hernández-Cassou, S., & Puignou, L., (2014). Characterization of fruit products by capillary zone electrophoresis and liquid chromatography using the compositional profiles of polyphenols: Application to authentication of natural extracts. *J. Agric. Food Chem., 62*(5), 1038–1046.

70. Rockenbach, I. I., Jungfer, E., Ritter, C., Santiago-Schübel, B., Thiele, B., Fett, R., & Galensa, R., (2012). Characterization of flavan-3-ols in seeds of grape pomace by CE, HPLC-DAD-MSn and LC-ESI-FTICR-MS. *Food Res. Int., 48*(2), 848–855.

71. Sturm, S., Seger, C., & Stuppner, H., (2007). Analysis of central European corydalis species by nonaqueous capillary electrophoresis-electrospray ion trap mass spectrometry. *J. Chromatogr. A, 1159*(1/2), 42–50.

72. Delatte, T. L., Schluepmann, H., Smeekens, S. C., De Jong, G. J., & Somsen, G. W., (2011). Capillary electrophoresis-mass spectrometry analysis of trehalose-6-phosphate in Arabidopsis thaliana seedlings. *Anal. Bioanal. Chem., 400*(4), 1137–1144.

73. Aebersold, R., & Goodlett, D. R., (2001). Mass spectrometry in proteomics. *Chem. Rev., 101*(2), 269–296.

74. Samskog, J., Bylund, D., Jacobsson, S. P., & Markides, K. E., (2003). Miniaturized on-line proteolysis-capillary liquid chromatography-mass spectrometry for peptide mapping of lactate dehydrogenase. *J. Chromatogr. A, 998*(1/2), 83–91.

75. Cobb, K. A., & Novotny, M., (1989). High-sensitivity peptide mapping by capillary zone electrophoresis and microcolumn liquid chromatography using immobilized trypsin for protein digestion. *Anal. Chem., 61*(20), 2226–2231.

76. Wang, C., Oleschuk, R., Ouchen, F., Li, J., Thibault, P., & Harrison, D. J., (2000). Integration of immobilized trypsin bead beds for protein digestion within a microfluidic chip incorporating capillary electrophoresis separations and an electrospray mass spectrometry interface. *Rapid Commun. Mass Spectrom., 14*(15), 1377–1383.

77. Křenková, J., Klepárník, K., & Foret, F., (2007). Capillary electrophoresis mass spectrometry coupling with immobilized enzyme electrospray capillaries. *J. Chromatogr. A, 1159*(1/2), 110–118.

78. Shen, X., Yang, Z., McCool, E. N., Lubeckyj, R. A., Chen, D., & Sun, L., (2019). Capillary zone electrophoresis-mass spectrometry for top-down proteomics. *Trends Anal. Chem.,* 115644.

79. Klen, T. J., Wondra, A. G., Vrhovsek, U., & Vodopivec, B. M., (2015). Phenolic profiling of olives and olive oil process-derived matrices using UPLC-DAD-ESI-QTOF-HRMS analysis. *J. Agric. Food Chem., 63*(15), 3859–3872.

80. Shao, Y.,Jiang, J., Ran, L.,Lu, C., Wei, C., & Wang, Y., (2014). Analysis of flavonoids and hydroxycinnamic acid derivatives in rapeseeds (*Brassica napus* L. var. *napus*) by HPLC-PDA-ESI (-)-MS$^n$/HRMS. *J. Agric. Food Chem., 62*(13), 2935–2945.

81. Lin, L. Z., Sun, J., Chen, P., & Harnly, J., (2011). UHPLC-PDA-ESI/HRMS/MS$^n$ analysis of anthocyanins, flavonol glycosides, and hydroxycinnamic acid derivatives in red mustard greens (*Brassica juncea* Coss Variety). *J. Agric. Food Chem., 59*(22), 12059–12072.

82. Qiao, X., Song, W., Wang, Q., Liu, K. D., Zhang, Z. X., Bo, T., Li, R. Y., et al., (2015). Comprehensive chemical analysis of triterpenoids and polysaccharides in the medicinal mushroom *Antrodia cinnamome. RSC Adv., 5*, 47040–47052.

83. El-Elimat, T., Figueroa, M., Ehrmann, B. M., Cech, N. B., Pearce, C. J., & Oberlies, N. H., (2013). High-resolution MS, MS/MS, and UV database of fungal secondary metabolites as a dereplication protocol for bioactive natural products. *J. Nat. Prod., 76*(9), 1709–1716.

84. Piccinelli, A. L., Pagano, I., Esposito, T., Mencherini, T., Porta, A., Petrone, A. M., Gazzerro, P., et al., (2016). HRMS profile of a hazelnut skin proanthocyanidin-rich fraction with antioxidant and anti-*Candida albicans* activities. *J. Agric. Food Chem., 64*(3), 585–595.

85. El-Elimat, T., Raja, H. A., Graf, T. N., Faeth, S. H., Cech, N. B., & Oberlies, N. H., (2014). Flavonolignans from *Aspergillusiizukae*, a fungal endophyte of milk thistle (*Silybum marianum*). *J. Nat. Prod., 77*, 193–199.

86. Sica, V. P., Raja, H. A., El-Elimat, T., Kertesz, V., Berkel, G. J. V., Pearce, C. J., & Oberlies, N. H., (2015). Dereplicating and spatial mapping of secondary metabolites from fungal cultures in Situ.*J. Nat. Prod., 78*(8), 1926–1936.

87. Oliveira, A. L., Destandau, E., Fougère, L., & Lafosse, M., (2014). Isolation by pressurised fluid extraction (PFE) and identification using CPC and HPLC/ESI/MS of phenolic compounds from Brazilian cherry seeds (*Eugenia uniflora* L.). *Food Chem., 145*, 522–529.

88. Riffault-Valois, L., Destandau, E., Pasquier, L., André, P., & Elfakir, C., (2016). Complementary food analytical methods for the phytochemical investigation of

'Jardin de Granville', a rose dedicated to cosmetics. *Comptes Rendus Chimie., 19*(9), 1101–1112.

89. Guo, H., Kreuzenbeck, N. B., Otani, S., Garcia-Altares, M., Dahse, H. M., Weigel, C., Aanen, D. K., et al., (2016). Pseudoxylallemycins A-F, cyclic tetrapeptides with rare allenyl modifications isolated from *Pseudoxylaria* sp. X802: A competitor of fungus-growing termite cultivars. *Org. Lett., 18*(14), 3338–3341.

90. Luo, J., Huang, W. S., Hu, S. M., Zhang, P. P., Zhou, X. W., Wang, X. B., Yang, M. H., et al., (2017). Rearranged limonoids with unique 6/5/6/5 tetra carbocyclic skeletons from *Toona ciliata* and biomimetic structure divergence. *Org. Chem. Front., 4*, 2417–2421.

91. Wu, J. W., Tang, C., Ke, C. Q., Yao, S., Liu, H. C., Lin, L. G., & Ye, Y., (2017). Dicarabrol A., dicarabrone C and dipulchellin A, unique sesquiterpene lactone dimers from carpesium abrotanoides, *RSC Adv., 7*, 4639–4644.

92. Zhang, W. Y., An, F. L., Zhou, M. M., Chen, M. H., Jian, K. L., Quasie, O., Yang, M. H., Luo, J., & Kong, L. Y., (2016). Limonoids with diverse frameworks from stem barks of *Entandrophragma angolense*and their bioactivities. *RSC Adv., 6*, 97160–97171.

93. Johansen, K. T., Wubshet, S. G., Nyberg, N. T., & Jaroszewski, J. W., (2011). From retrospective assessment to prospective decisions in natural product isolation: HPLC-SPE-NMR analysis of *Carthamus oxyacantha. J. Nat. Prod., 74*, 2454–2461.

94. Seger, C., Godejohann, M., Tseng, L. H., Spraul, M., Girtler, A., Sturm, S., & Stuppner, H., (2005). LC-DAD-MS/SPE-NMR hyphenation. A tool for the analysis of pharmaceutically used plant extracts: Identification of isobaric iridoid glycoside regioisomers from *Harpagophytum procumbens.Anal. Chem., 77*(3), 878–885.

95. Gurmessa, G. T., Kusari, S., Laatsch, H., Bojase, G., Tatolo, G., Masesane, I. B., Spiteller, M., & Majinda, R. R. T., (2018). Chemical constituents of root and stem bark of *Erythrina brucei. Phytochem. Lett., 25*, 37–42.

96. Han, J., Niu, S. T., Liu, Y., Gan, L., wang, T., Lu, C. D., & Yuan, T., (2018). Robustanoids A and B, two novel pyrrolo[2,3-b]indole alkaloids from *Coffea canephora*: Isolation and total synthesis. *Org. Chem. Front., 5*, 586–589.

97. Kato, L., De Oliveira, C. M., Melo, M. P., Freitas, C. S., Schuquel, I. T., & Delprete, P. G., (2012). *Glucosidic iridoids* from *Molopanthera paniculata* Turcz. (Rubiaceae, Posoquerieae). *Phytochem. Lett., 5*(1), 155–157.

98. Zhang, X., Han, B., Feng, Z., Yang, Y. N., Jiang, J. S., & Zhang, P. C., (2017). Phthalide derivatives from *Ligusticum chuanxiong. RSC Adv., 7*, 37478–37486.

99. Yang, Y. N., An, Y. W., Zhan, Z. L., Xie, J., Jiang, J. S., Feng, Z. M., Yea, F., & Zhang, P. C., (2017). Nine new compounds from the root bark of *Lycium chinense* and their a-glucosidase inhibitory activity. *RSC Adv., 7*, 805–812.

100. Mahar, R., Dixit, S., Joshi, T., Kanojiya, S., Mishra, D. K., Konwar, R., & Shukla, S. K., (2016). Bioactivity guided isolation of oxypregnane-oligoglycosides (Calotroposides) from the root bark of *Calotropis gigantea* as potent anticancer agents. *RSC Adv., 6*, 104215–104226.

101. Xiong, J., Hong, Z. L., Xu, P., Zou, Y., Yu, S. B., Yang, G. X., & Hu, J. F., (2016). Ent-Abietane diterpenoids with anti-neuroinflammatory activity from the rare *Chloranthaceae* plant *Chloranthusoldhamii. Org. Biomol. Chem., 14*, 4678–4689.

102. Thai, Q. D., Tchoumtchoua, J., Makropoulou, M., Boulaka, A., Meligova, A. K., Mitsiou, D. J., Mitakou, S., et al., (2016). Phytochemical study and biological evaluation of chemical constituents of *Platanus* oriental is and platanus × acerifolia buds. *Phytochemistry, 130*, 170–181.

103. Saidi, I., Waffo-Téguo, P., Ayeb-Zakhama, A. E., Harzallah-Skhiri, F., Marchal, A., & Jannet, H. B., (2018). Phytochemical study of the trunk bark of *Citharexylum spinosum* L. growing in Tunisia: Isolation and structure elucidation of iridoid glycosides. *Phytochemistry, 146*, 47–55.

104. Hulcová, D., Maříková, J., Korábečný, J., Hošťálková, A., Jun, D., Kuneš, J., Chlebek, J., et al., (2019). Amaryllidaceae alkaloids from *Narcissus pseudonarcissus* L. cv. Dutch master as potential drugs in treatment of Alzheimer's disease. *Phytochemistry, 165*, 112055.

105. Tsafantakis, N., Zelianeos, K., Termentzi, A., Vontzalidou, A., Aligiannis, N., & Fokialakis, N., (2019). Triterpenes from *Echinopsspinosissimus* Turra subsp. spinosissimus. *Phytochem. Lett., 30*, 273–277.

106. Choudhary, A., Kumar, R., Srivastava, R. B., Surapaneni, S. K., Tikoo, K., & Singh, I. P., (2015). Isolation and characterization of phenolic compounds from *Rhodiola imbricata*, a Trans-Himalayan food crop having antioxidant and anticancer potential. *J. Funct. Foods, 16*, 183–193.

107. Bedane, K. G., Kusari, S., Masesane, I. B., Spiteller, M., & Majinda, R. R., (2016). Flavanones of *Erythrinalivingstoniana* with antioxidant properties. *Fitoterapia, 108*, 48–54.

108. Bedane, K. G., Masesane, I. B., & Majinda, R. R., (2016). New isoflavans from the root bark of *Erythrinalivingstoniana*. *Phytochem. Lett., 17*, 55–58.

109. Gurmessa, G. T., Kusari, S., Laatsch, H., Bojase, G., Tatolo, G., Masesane, I. B., Spiteller, M., & Majinda, R. R., (2018). Chemical constituents of root and stem bark of *Erythrina brucei*. *Phytochem. Lett., 25*, 37–42.

110. Riffault, L., Destandau, E., Pasquier, L., André, P., & Elfakir, C., (2014). Phytochemical analysis of *Rosahybrida* cv. 'Jardin de Granville' by HPTLC, HPLC-DAD, and HPLC-ESI-HRMS: Polyphenolic fingerprints of six plant organs. *Phytochemistry, 99*, 127–134.

111. Tuenter, E., Foubert, K., Staerk, D., Apers, S., & Pieters, L., (2017). Isolation and structure elucidation of cyclopeptide alkaloids from *Ziziphus nummularia* and *Ziziphus spina-christi* by HPLC-DAD-MS and HPLC-PDA-(HRMS)-SPE-NMR. *Phytochemistry, 138*, 163–169.

112. Zeraik, M. L., Queiroz, E. F., Marcourt, L., Ciclet, O., Castro-Gamboa, I., Silva, D. H. S., Cuendet, M., et al., (2016). Antioxidants, quinone reductase inducers, and acetylcholinesterase inhibitors from *Spondias tuberosa* fruits. *J. Funct. Foods, 21*, 396–405.

113. Cerulli, A., Napolitano, A., Masullo, M., Hošek, J., Pizza, C., & Piacente, S., (2020). Chestnut shells (Italian cultivar "Marrone di Roccadaspide" PGI): Antioxidant activity and chemical investigation with in depth LC-HRMS/MS[n] rationalization of tannins. *Food Res. Int., 129*, 108787.

114. Nikolaivits, E., Termentzi, A., Skaltsounis, A. L., Fokialakis, N., & Topakas, E., (2017). Enzymatic tailoring of oleuropein from Olea europaea leaves and product identification by HRMS/MS spectrometry. *J. Biotechnol., 253*, 48–54.

115. Kee, C. L., Chin, L. C., Cheah, N. P., Ge, X., & Low, M. Y., (2019). Elucidation of the absolute configuration of a tadalafil analog found as adulterant in a health supplement by mass spectrometry, chiroptical methods, and NMR spectroscopy. *J. Pharm. Biomed. Anal., 173*, 47–55.

116. Kamtcha, D. W., Tene, M., Bedane, K. G., Knauer, L., Strohmann, C., Tane, P., Kusari, S., & Spiteller, M., (2018). Cardenolides from the stem bark of *Salaciastaudtiana. Fitoterapia., 127*, 402–409.

117. Tebou, P. L. F., Ngnokam, D., Harakat, D., & Voutquenne-Nazabadioko, L., (2018). Three new iridolactone derivatives from the whole plant of *Brillantaisiaowariensis* P. Beauv. *Phytochem. Lett., 25*, 171–174.

118. Plazas, E., Casoti, R., Murillo, M. A., Da Costa, F. B., & Cuca, L. E., (2019). Metabolomic profiling of *Zanthoxylum* species: Identification of anti-cholinesterase alkaloids candidates. *Phytochemistry, 168*, 112128.

# CHAPTER 5

# High-Resolution Bioassays as Preparation Screening Techniques

SHINTU JUDE and SREERAJ GOPI

*Research and Development (R&D) Center, Plant Lipids (P) Ltd., Kadayiruppu, Kolenchery, Cochin, Ernakulam, Kerala – 682311, India*

## ABSTRACT

In any field that deals with phytochemicals, their characterization is an important part as the basic factor is to achieve a bioactive compound or an extract rich in bioactive components. Therefore, the procedures indulged in the same were supposed to support the aim. Bioassays are meant to find out the effect of analyte on biological systems and hence deliver direct results on the bioactivity of the extract of interest, which allows the rapid fractionation and purification of the natural product extract. In addition to the fractionation procedure, it serves well for providing a first-line view about the activity nature also. In phytochemical characterization using HRMS, bioassays serves as a preparatory technique, which is detailed in the chapter.

## 5.1 INTRODUCTION

Mother Nature provides remedies for many of our troubles, and the related questions are well answered by plant secondary metabolites. Many attempts have been made to develop effective methods for the investigation of potentially bioactive molecules from herbal sources and have resulted in the form of products, methods, technologies, and many more. One important category among them is bioassays. With many add-ons invented later, they present themselves as rapid, low cost, and streamlined approaches for bioactive compounds determination.

Bioassays can be defined basically as the analytical procedure to determine the potency of any substance on biological systems such as antigen-antibody reactions, living cells or tissues, living animals, agricultural systems, etc. Here, the efficacy of the substance against the biological system is measured as equivalents of some standards. Bioassays are classified under various categories depending on different peculiarities such as sample size and nature, experimental design, target life forms, and the mode of response produced [1]. It can be carried out in different systems, ranging from subcellular systems to whole animals. Considering all these natures together, bioassays are classified into two primary and secondary. Primary bioassays allow the testing in a number of samples within short time and provide basic information on the activity of the sample of interest. The lead compounds obtained from the primary bioassay are investigated in detail, in order to produce specific and comprehensive data and the process is termed as secondary bioassays.

Bioassays can be tailored for many purposes depending on the conditions and requirements. They can act as guiding procedures and can be the major part of the isolation processes to isolate the active components. Anyhow, in the new era, this pave way to many doors such as drug designing and development, quantification, and quality assurance of active compounds. Bioassays are found to be easy as well as rapid and have been tooled as the mode of exploring pharma activities such as active compound isolation and identification. Here, the sample is primarily fractionated by some separation techniques, and the fractions are submitted for bioassay screening to distinguish both the active components and their activities. The introduction of different types of chromatography allows an open system to conduct such screenings in an effortless manner. By further hyphenations with HRMS, they furnish a strong team, so as to elucidate the structure and identify the active compounds. To justify the title, the chapter discusses different bioassays working well in phytochemicals.

The procedure for bioassay analyzes consists of an analytical scale HPLC elution and the micro-fractionation of these eluates (as a function of time) into microplates, followed by bio-assaying of the content of each well. A biochromatogram will result, which can easily correlate with the HPLC chromatogram and can match the bioactivity of each component. The coupling of fractionation and bioassays can be of either way-it can be a bioactivity guided fractionation, or a fractionation followed by bioactivity investigation. A biochromatogram, along with databases or hyphenated techniques such as NMR, HRMS, etc., can serve as a perfect library to identify the unknown compounds.

## 5.2 BIOASSAYS AS SCREENING TECHNIQUES

Let the name be dereplication, high throughput screening or whatever, all the analytical procedures involved in the processes are aimed towards the investigation of bioactive ingredients in the extract. It is not advisable to introduce the crude extract directly to the analytical instrument, as the matrix may interfere in our concerns. So, bioanalytical screening techniques have been introduced as a remedy to reduce the intrusion of complex crude matrix in the active component characterization. Many researchers have conducted a pre-purification of the extract, which could have an impact on the number of false identifications or undetected minor bioactive compounds. This step, if preceded with a 'target guided' manner, can accelerate the process in a positive way. In addition, the quality and bioactivity of natural products are ensured by bioassays in many studies. Intelligent use of biomarkers allows effective target identification too. Many researchers make use of high-resolution bioassay prior to the hyphenated analysis setup "HPLC-HRMS-SPE-NMR," in order to discern the individual components' pharmacological activity and then to derive the structure. This coupling of instruments has proven to be very effective that, in the first step, the bioactive fractions will be chosen and in the second step, the characterization is conducted only for the selected fractions. This one additional step could cut down the tedious time consuming screening steps and allows a simultaneous biological and chemical screening of extracts. A schematic representation is given in Figure 5.1.

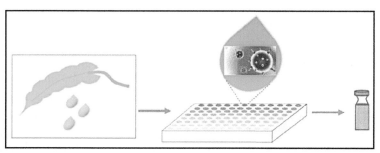

**FIGURE 5.1** Bioassay: high throughput screening, where, a tray of minimum 96 wells is used for monitoring the reactions and responses of the test samples against assay reagents.

## 5.3   BIOASSAYS IN CURE FOR LIFE STYLE DISEASES

Nowadays, the number of sufferers of lifestyle diseases is increasing like anything. Type 2 diabetes acquires more attention among them, as the progression of disease is very rapid and risky as it causes secondary comorbidities. Modern medicines for type 2 diabetes have found to cause some side effects also [2–5]. At this point, natural remedies for the disease got much recognition and many studies have investigated for the antidiabetic constituents from nature. Most of the natural antidiabetic constituents possess the ability to suppress the conversion of carbohydrates into glucose in the gut, or to inhibit the absorption of glucose from the intestine. However, the real hurdle is to isolate the active compounds from the complex natural herbal matrices. A simple and advanced way for this is to use bioassays such as α-glucosidase (AGH) inhibition assay, protein-tyrosine phosphatase 1B (PTP1B) inhibition assay-amylase inhibition assay, etc., in order to separate the active constituents from the crude extract. A major consideration of high-resolution bioassays in herbal therapeutics is the separation, identification, and structure elucidation of active compounds. Hence, much of these assays are utilized to pilot the hyphenated analysis layout HPLC-HRMS-SPE-NMR.

### 5.3.1   ALPHA-GLUCOSIDASE (α-GLUCOSIDASE) INHIBITION ASSAYS

α-glucosidase (AGH), present in the enterocytes of jejunum is a key enzyme, which catalyzes the hydrolysis of starch, oligosaccharides, and disaccharides into monosaccharides and enhances the absorption of glucose. An AGH inhibitor can delay these activities and can downregulate the absorption of glucose and can control the after meal insulin levels. As the AGH inhibitors can serve as the first line therapy for diabetes, their natural resources are studied extensively. One easy, straight cut and efficient strategy for this is to use AGH inhibitor bioassay. These bioassays allow the simple, rapid isolation of AGH inhibitors distributed all around the globe.

The proven incidents of AGH inhibitory activities of phytochemicals delivered the provisions for utilization of bioassays in this regard. One of the first studies has used bioassays hyphenated with HPLC-SPE-NMR for AGH inhibitors identification, by using apple peel as an example.

Five compounds were identified to be potential, among which, one compound-reynoutrin has no precedent studies of AGH inhibition activity [6]. HPTLC-AGH inhibitor bioassay was successfully used for the identification of AGH inhibitor compounds from cherimoya fruit (*C. cherimola* Mill.). Three compounds were found to be AGH inhibitors and were chemically verified as phenolamides, among which, one was identified for the first time from the fruit [7]. A bioassay guided AGH inhibitory activity study was conducted for the stem of *Vigna angularis*. Sixteen compounds were found to be the bioactive components and they were proceeded for structural characterization by NMR and HR-ESI-MS data. Among the identified bio-actives, one was a new compound and another one was found in a natural source for the first time. Along with the activity of separated components, the study provided structure-activity relationships (SARs) of each inhibitor with AGH from molecular docking study as well [8]. Leaves of *Clinacanthus nutans* are rich source of potential inhibitors with assured AGH inhibitory activity, as evidenced by the bioassays. The compounds, acted as AGH inhibitors were identified as "N-Isobutyl-2-nonen-6,8-diynamide," "tabanone," "1,'2'-bis(acetyloxy)-3,'4'-didehydro-2'-hydro-β, ψ-carotene" and "22-acetate-3-hydroxy-21-(6-methyl-2,4-octadienoate)-olean-12-en-28-oic acid." While the *in vitro* assay proved the bioactivity of compounds, *in silico* technique provided evidence for their synergic activity and molecular docking provided binding modes of these compounds with ABH [9].

Eder Lana e Silva and coworkers identified six AGH inhibitors and nineteen other metabolites from the leaves of *Eremanthus crotonoides* by high-resolution AGH inhibition profiling combined with HPLC-HRMS-SPE-NMR, seconding the importance of the species as a remedy for type 2 diabetes [10]. Four compounds from the leaves of *Crataegus oxyacantha* L. (Hawthorn) were identified as AGH inhibitors with LC-DAD-ESI-MSn and structures of the same were affirmed by sustained off-resonance irradiation collision-induced dissociation (SORI-CID) data from FT-ICR-MS [11]. Twenty-two lanostane-type triterpenoids showing AGH inhibitory activity was isolated from the fruiting bodies of *Ganoderma hainanense,* by using *in vitro* α-glucosidase inhibitory assay, whose structures were elucidated by HR-ESI-MS-IR-NMR, and their SAR was established [12].

Research for bioactive compounds takes man to anywhere, even into the deeps. AGH inhibitory properties of sea aster (*Aster tripolium* L.) and searocket (*Cakile maritima* Scop.) were studied using high-resolution

AGH inhibition assays followed by HPLC-HRMS-SPE-tube transfer NMR. Three AGH inhibitors were identified from sea aster and structural characterization was done by asserting micrOToF-Q II MS in the hyphenation [13]. The same analysis framework was used to analyze a total of 57 crude extracts from 19 different edible seaweeds for their AGH inhibitory constituents. Two classes of bioactive compounds, phlorotannins and fatty acids, were identified to be present in the extracts from brown seaweed, which showed the higher AGH inhibitory activity among all the analyzed [14]. *Myrcia palustris* DC. (Myrtaceae), a renowned traditional antidiabetic agent from Brazilian forests didn't have much documented history on its active constituents. In 2015, Wubshet et al. reported the isolation and characterization of AGH inhibitors from the leaves of the plant, by employing AGH bioassay in series with HPLC-HRMS-SPE-NMR. Out of the 20 major compounds, five were evidenced as AGH inhibitors, including two well known AGH inhibiting flavonoid myricetin and quercetin, along with three unprecedented inhibitors-casuarinin, myricetin 3-O-β-D-(6"-galloyl)galactopyranoside, and kaempferol 3-O-β-D-galactopyranoside [15].

A synergic potential of a combination of two different bioactivity profiling techniques-high-resoultion bioassay and ligand fishing (please refer Chapter 6) was used for the characterization of *Ginkgo biloba* extract. Both the techniques acted complimentary and resulted in reduced false positive identifications. HRMS-NMR platform rendered the identification of all the AGH inhibitors as bioflavonoids [16].

### 5.3.2   PTP1B ASSAY

Insulin-signaling pathway plays a key role in the cell glucose regulation. Ligand binding with insulin receptor causes autophosphorylation of the same, which in turn triggers the signaling pathway. As a result, type 4 glucose transporters are upregulated and leads to the increase in glucose uptake from the blood. PTP1B is an enzyme, which could dephosphorylate the insulin receptor and can, thus block the signaling initiation. Thus, PTP1B holds a position of negative regulator in the insulin-signaling pathway. They also function as the negative regulator in the signaling of leptin, which is an important modulator of energy homeostasis and body weight. Altogether, PTP1B outfits the role of "regulator" in many influential conditions such as diabetes, obesity, neuroinflammation, Alzheimer's disease (AD), etc., also, the PTP1B knockout studies demonstrated an improved insulin sensitivity and reduced diet induced obesity, seconding the aforementioned effects.

In a study, three species of *Eremophila* were checked for their antidiabetic components with the help of micrOToF-Q II. The work used high-resolution AGH and high-resolution PTP1B inhibition profiling followed by HPLC-HRMS-SPE-NMR for the screening of *Eremophila gibbosa*, *E. glabra*, and *E. aff. drummondii* and identified a total of 21 compounds, including 12 tanes [17]. The same hyphenated analysis series, including Human recombinant PTP1B enzyme was employed by Tahtah et al. and successfully pinpointed the diterpene named 5-hydroxyviscida-3,14-dien-20-oic acid as responsible for the antidiabetic features of *Eremophila lucida* [18]. Zhao et al. searched for the antidiabetic properties of the stem and root of *Polygonum cuspidatum* Siebold and Zucc. By following a dual high-resolution AGH and PTP1B inhibition bioactivity profiling in combination with HPLC-HRMS-NMR. The analysis setup of two stage bioactivity profiling enabled the isolation of pharmacologically active compounds easier. They have identified the structural features of 16 compounds responsible for the antidiabetic activity of the crude extract of *P. cuspidatum* by using micrOToF-Q II mass spectrometer as a part of the analysis setup [19]. In a similar way, number of compounds from *Eremophila bignoniiflora* were identified as potential PTP1B inhibitors by employing bioassay profiling and pharmacological evaluation. Here, a simple combination of LC-HRMS and NMR could identify these active compounds like flavonoids and furanone sesquiterpenes [20].

### 5.3.3  5'ADENOSINE MONOPHOSPHATE-ACTIVATED PROTEIN KINASE (AMPK) INHIBITION ASSAY

AMPK is related to the regulation of carbohydrate and glucose metabolism, since its activation triggers ATP production. Hence, AMPK inhibition is used as a fishing property for antidiabetic constituents. A commercial extract of *Lippia citriodora* was characterized for its phenolic profile. The extract was fractionated and subjected to AMPK modulation study utilizing 3T3-L1 adipocyte model. 29 active compounds including iridoids, phenylpropanoids, and flavonoids were identified from 11 fractions by RP-HPLC-ESI-ToF-MS [21].

### 5.3.4　ALPHA AMYLASE (α-AMYLASE)

The enzyme α-amylase plays a role in the cleavage of insoluble large carbohydrate molecules into smaller, soluble carbohydrates and inside the body, they serve significant strategies for blood glucose level management. Antidiabetic activity of cinnamon (*Cinnamomum verum* Presl.) was studied by applying α-amylase inhibition assay together with HPLC-HRMS-SPE-NMR. Both the chemical and biological profiling of cinnamon was conducted, proving that cinnamaldehyde is the major α-amylase inhibitor along with four other minor compounds [22]. A quadrupole high-resolution AGH, PTP1B, α-amylase, radical scavenging profiling followed by activity guided analysis with HPLC-HRMS-SPE-NMR was carried out for *Morus alba* L. root bark extract. A number of Diels-Alder adducts, including a new one named Moracenin E along with other isoprenylated flavonoids and 2-arylbenzofurans were identified as the bioactive components and they were found to have high radical scavenging activity also [23].

### 5.3.5　ACETYLCHOLINESTERASE (ACHE)

Different types of Dementia, especially AD, are becoming a major health challenge as it affects the quality of life experienced, both personally and socially. AD is characterized by a number of abnormalities connected to memory, behavior, thinking, language, and many other related functions, assumed to occur due to a deficit of cholinergic functions in the brain. Acetylcholine is a neurotransmitter, which is an active part of cholinergic synapses and deals with the cognitive functions. Acetylcholinesterase (AChE) hydrolyzes acetylcholine and thus creates a hurdle for the activity thereof. Therefore, one of the most important approaches in the therapy of AD is regulating the acetylcholine levels by administering acetylcholinesterase inhibitors (AChEIs) and AChEIs are considered as the widely accepted remedy for AD. Natural AChEI sources are of great importance in this scenario. A mass study in this area was conducted in Iran, in which, AChE inhibitory effects were evaluated for 25 Iranian plants. A combination of AChE inhibition assay NMR-HRMS-Molecular networks-molecular docking have suggested the n-hexane extract of *Prangosferulacea* as the most effective AChE inhibitor, owing to the presence of many active constituents, including flavonoids and furanocoumarins [24].

## 5.3.6 ANDROGEN BIOASSAY

Androgens are hormones; participate in many body functions such as metabolism, insulin sensitivity, bone, and cardiovascular health, etc., and most importantly, in sexual health and reproductive system [25]. A notable study was the development of an identification procedure to separate and identify unknown androgen, pro, androgen, and androgen derivatives in herbal preparations and sports supplements. By including three different pretreatment steps, it was made possible to activate the inactive compounds also. A yeast androgen bioassay-guided fractionation, followed by a simple combination of UHPLC/ToFMS and an accurate mass database (MetAlign) was successfully employed for compound identification [26].

## 5.3.7 LIPOXYGENASE (LOX) INHIBITION ASSAY

The enzyme lipoxygenase (LOX) catalyzes the arachidonic acid metabolism by specifically inserting molecular oxygen in the molecule to form 5-hydroxyeicosatetraenoic acid, which in turn dehydrated to an unstable intermediate leukotriene (LT) A4, which is metabolized to cysteinyl leukotrienes (CysLTs) by LTC4 synthase. Thus, the LOX inhibitors furnish the therapeutic tool for many chronic inflammatory diseases and Melicope ptelefolia Champ ex Benth was investigated for the same. By using inhibition towards soybean 5-LOX activity as a tool, the fractionation was carried out and subjected to test inhibitory effects against formation of cysLTs and human PBML 5-LOX. 2,4,6-trihydroxy-3-geranylacetophenone (tHGA) was found to be the active compound and a dual mechanism of action proposed [27].

## 5.4 ANTIOXIDANT ASSAYS

Antioxidant assays do not impart any direct biological processes, but are considered to generate biologically relevant results and hence assays corresponding to antioxidant and free radical scavenging is classified under bioassays [28].

An antioxidant bioassay was conducted for the aqueous ethanol extract of roots of *Solanum melongena,* in terms of nitric oxide (NO) production induced by lipopolysaccharides in e RAW 264.7 cell line. Three new lignanamides, along with 14 known analogs were found to be the NO inhibitory compounds in the extract. A combination of semi-preparative HPLC-TLC-NMR-HRMS was utilized for the structure elucidation of the identified compounds. The cell viability study by MTT assay, conducted prior to the bioassay confirmed the absence of cytotoxicity of the isolates, supporting their inhibitory activity against NO production [29]. *Artocarpus heterophyllus* L. commonly known as jackfruit is renowned for its tasty edible pulp of ripe fruits, medicinal value of raw fruits and for having the world record of biggest fruit. This fruit produces a big quantity of fruit waste, and this fruit waste was studied for its antioxidant potential. J33 variety was considered for the study and its ethyl acetate fraction was subjected to the bioassay fractionation by using DPPH assay. Chemical constituents were identified by ToF-LCMS with the help of data analyzer-mass hunter and an online database. Fifteen active constituents were found to contribute to the antioxidant activity of the extract out of which, 11 compounds were identifying from the particular species for the first time [30]. Dash et al. conducted a complete study on the strong antioxidant-Pentylcurcumene-from *Geophila repens* (L.) I. M. Johnst (Rubiaceae). Along with the HPTLC fingerprinting, cellular antioxidant protocols such as DPPH assay, oxygen radical absorbance capacity (ORAC) assay and cell-based antioxidant protection in erythrocytes (CAP-e) assay were performed to ensure the antioxidant activities [31].

*Salvia miltiorrhiza* Bunge root (Danshen) is a popular traditional medicine for its beneficiary properties against bone loss, cancer, renal deficiency, Alzheimer's, Parkinson's, cerebrovascular, coronary heart diseases, hepatocirrhosis, and skin problems. Ethyl acetate extract of the herb was subjected to an antioxidant response element (ARE)-driven luciferase reporter system. The strategy was to remove the major active compounds by high-resolution peak fraction approach and to enhance the prospective discovery of active components in trace amounts by bioassay-HRMS parallel coupling. Among a total of 62 minor compounds, 33 were found to be nuclear factor erythroid 2-related factor 2 (Nrf2) activating molecules and thus are important in treatment of oxidative stress and related diseases [32].

## 5.5  ANTI-MICROBIAL ACTIVITY

Microbial contamination is one major reason behind the triggering and propagation of many of the diseases and hence the antimicrobial compounds, especially natural compounds find their importance as thera-peutic substances. We are discussing about antibacterial, antifungal, and anthelmintic compounds.

A study illustrated a rapid and efficient analysis method using high-resolution anti-bacterial profiling followed by HPLC-HRMS-SPE-NMR for the screening of 180 extracts from 88 traditional Chinese plant species, which have been used as folk medicine for snakebite. The antibacterial potential of the plants was investigated as the necrotic wound, caused by snakebite can secondarily be affected by a bacterial infection from the snakemouth itself. The hybrid of instruments allowed a fast pinpointing from a complex matrix and the bio-chromatograms revealed that tannins are the major bacterial growth inhibitors in the studied plants, along with three non-tannin active compounds [33]. The presence and activities of tannins against snakebite necrosis have proved by a high-resolution hyaluronidase inhibition profiling accompanied with HPLC-HRMS-SPE-NMR analysis. Eighty-eight plant species, traditionally used to cure snakebite in China, were investigated for their hyaluronidase, phospho-lipase A2 and protease inhibitor activities against four different venoms. Among them, 22 plant extracts, due to their high content of tannins, found to have a good activity against at least one venom. In addition, the study allowed structural identification of four active non-tannin components [34]. *Aliivibrio fischeri* bacteria suspension-bioassay was employed well as another important tool for the characterization of active compounds from the berry extracts of bilberry, blueberry, chokeberry, açai berry, and cranberry. HPTLC-Vis MS served even as a quantitative aid and the study put forward a rapid and simple quality assessment procedure, including a DPPH assay, direct antibacterial bioassay, and HRMS [35]. *Tanacetum vulgare L.* (tansy) root extract was subjected to separation by overpres-sured layer chromatography (OPLC)-DART-HRMS, which accomplished a rapid and efficient separating tool. Bioassays using *A. fischeri* and *B. subtilis* along with the OPLC-DART-HRMS hyphenation could isolate six active compounds, including four polyacetylene compounds [36].

Due to the anthelmintic resistance, the parasites may survive the drug doses and often leads to the failure of anthelmintic drugs. It raises a requisite of newer drugs and doses. *Warburgia ugandensis,* a Sprague subspecies *ugandensis* leaves were studied for their anthelmintic proper-ties, using N2 wild-type *C. elegans* strain, followed by cytotoxicity assay using HEK 293 and RAW 264.7 cell lines. Three active components were identified and characterized as warburganal, polygodial, and ALA. Apart from being an active component, individually warburganal and polygodial are synergistic with ALA. According to anthelmintic activity and cytotoxicity, polygodial was the most potent compound and was selected for further investigations. Interestingly, polygodial found to block the *C. elegans* motility by a specific mechanism than other compared, established anthelmintic drugs, as the *C. elegans* mutant strains could not show resistance against polygodial. Besides, polygodial exhibits considerable inhibition towards mitochondrial ATP synthesis of *C. elegans* with respect to the dose. But, as it is shown in a SAR, the efficiency of polygodial rely on the α,β-unsaturated 1,4-dialdehyde structural motif [37].

Natural fungicides are comparatively new vision in product categorization. Plants are known to possess antifungal chemical moieties in them, and the efforts for isolating the potent antifungal agents are seeking attention of the time. Highly specified fungicides are developed by identifying and enhancing the plasma membrane (PM) H⁺-ATPase inhibitors. A research on *Candida albicans* along with *Saccharomyces cerevisiae* targeted the PM H⁺-ATPase enzyme from the fungi as the center of action for antifungal activity of the plant bio-actives. Around 46 crude extracts from 33 different plant species have screened by HPLC-HRMS-SPE-NMR analysis setup and the potent PM H⁺-ATPase inhibitors-chebulagic acid and tellimagrandin II were isolated and characterized from *Haplocoelum foliolosum* [38]. A high-resolution fungal (PM) H⁺-ATPase inhibition screening followed by HPLC-HRMS-SPE-NMR was used by Kongstad and coworkers for the identification of antifungal compounds from *Uvaria chamae* P. Beauv. A number of o-hydroxybenzylated flavanones and chalcones were identified in the study [39].

Leishmaniasis is a disease caused by parasites of almost 20 species of *Leishmania* and is considered as a neglected tropical disease, though it causes even death. Its treatment is very rare and expensive. A research for natural remedy for this health issue pave the way for a semi-high-resolution antileishmanial inhibition profiling combined with HPLC-HRMS-SPE-NMR for the isolation and characterization of antileishmanial compounds from leaves of *Lawsonia inermis* L. (henna) [40]. Five components were found to have inhibitory effect on *Leishmania tropica*, with two of them named luteolin and lalioside being excellent inhibitors.

The applications of bioassay-HRMS combination have an extended space in separating out the toxic compounds from the extracts. One major finding among the same was the nephrotoxic constituents recognized from the complex mixture of compounds extracted from Easter lilies (*Lilium longiflorum*). Feline epithelial kidney cell line CCl-94 was used against a series of three cytotoxicity assays-LDH leakage, Alamar blue reduction, and Neutral red uptake. The separated compounds were subjected to multiple fragmentation ion trap (IT) and HRMS profiling by using Accurate Mass Q/ToF mass spectrometer (MS) and identified to be steroidal glycoalkaloids (SGA) of solasodine nature. Different SGA analogs in the plant were identified as the major cytotoxic components of the Easter lily flowers extracts [41].

## 5.6   LARVAL BIOASSAYS

The essential oil from aerial part of *Adenosma Buchneroides* was investigated for the compounds responsible for its mosquito repellent activity. The in-cage mosquito repellent bioassay guided isolation of active compounds was further extrapolated to the identification by GC-MS and GC-FID, along with the structural elucidation by IR, HR-ESI-MS, and NMR. Again, the larval bioassays and MTS assays were employed to evaluate the larvicidal activity and cytotoxicity of these compounds. Thus the study furnishes the intelligent utilization of two related bioassays for two different intentions/ applications [42].

## 5.7   ANTICANCER ASSAYS

High potent and specific anticancer drugs are of great consideration, as the nonspecific anticancer drugs cannot distinguish between the normal cells and affected cell, and causes apoptotic cell death for both as well, which leads to many side effects and loss of immunity. Caspase-3-a cysteine protease which is specifically activated by apoptotic inducers, and then specifically inactivates a number of cellular proteins, causing cell death-is used in many high throughput-screening processes. Bioassay profiling of Gamboge, the resin of *Garcinia hanburyi*, including HeLa-C3 against caspase-3 could identify two potential fractions from the extract. Compound identification and characterization using HPLC/ESI Q-ToF MS ended up in gambogenic acid, epimeric isogambogic acids, and epimeric mixtures of gambogic acids (68.7%) and gambogoic acids (26.9%) [43]. Compounds from the leaves of *Macaranga barteri* were assessed for their cytotoxicity against human adenocarcinoma cell lines such as breast (MCF7), cervix (HeLa), lung (A549) and prostate (PC3), post cell viability studies. A total of 9 new compounds, 6 stilbenes, and 3 flavonols were isolated, out of which, stilbenes exhibited potent cytotoxicity, whereas flavonols showed a lesser potential. The measured selectivity index against human prostate cells proved the compounds' specific cytotoxicity towards the cancer cells [44].

Many of the phytochemical compounds act more effectively while in a combined form. This synergic effect is lost during partition and the individual compounds may less effective than complex crudes. The cytotoxicity effect of *Raphanus sativus* var. *caudatus* Alef onv against HCT116 cells was found to be attributed to many compounds including glucosinolates (GSL), isothiocyanates, indoles, flavonoids, alkaloids, thiocyanates, oxazolidine, and dialkyl disulfide and this slice of information allows the authentication and quality control (QC) of the product in laboratory scale, leaving the door of bulk quality assurance opened. UHPLC/ESI-QToF-MS/MS could deliver more detailed data than a similar previous study using GC-MS [45, 46].

## 5.8  COMBINATION BIOASSAYS

Combinations bioassays provide complementary therapeutic effects and promote the investigation of synergic activity of the bioactive compounds contribute towards effective therapeutic activities. Therefore, the compounds separated by combination bioassays will be more potent. It has been previously observed that geranylated flavonoids contribute to the bioactivity of *Paulownia tomentosa* (Thunb.) Siebold and Zucc. ex Steud in terms of antimicrobial, cytotoxic, and antioxidant effects. The potential inhibition raised by the fruit extract against AGH and PTP1B enzymes were used for the activity-based purification of compounds. Those compounds which have been proved as potent were subjected to structure elucidation and enzyme kinetics study [47].

In a study published in 2014, the crude methanol extract of a perennial leguminous vine *Pueraria lobate*-one among the 50 fundamental herbs in Traditional Chinese Medicine (TCM)-was screened by a promising combination of dual high-resolution AGH inhibition and radical scavenging profiling hyphenated with HPLC-HRMS-SPE-NMR. The study successfully presented a bioactivity profiling of the extract, where 21 compounds were of proven activity and three were new compounds, which were subjected to structure elucidation. The entire compound was proved to be potent AGH inhibitors and antioxidants [48].

A combination of three different high-resolution biological activity assays such as radical scavenging, α-glucosidase inhibition and aldose reductase inhibition assay was applied for the evaluation of Radix Scutellariae, the dried root of *Scutellaria baicalensis*. The triple high-resolution bioassay, in fusion with HPLC-micrOToF-Q II MS-SPE-NMR experiments in the crude extract allowed the identification of two major aldose reductase inhibitors, one α-glucosidase inhibitor and eight main radical scavengers [49]. Another effectively used triple combination was "high-resolution radical scavenging/α-glucosidase/α-amylase profiling" for the identification of bioactive compounds from *Dendrobium officinale*, followed by the HPLC-PDA-HRMS-SPE-NMR profiling where 6 AGH inhibitors, 1 alpha amylase inhibitor, and 12 radical scavengers were identified from the extract [42].

The pancreatic cholesterol esterase enzyme is an important enzyme which could upregulate the serum cholesterol levels. Angiotensin converting enzyme (ACE) takes part in the modulation of body fluid volume and thereby involved in the regulation of blood pressure levels. So, the inhibitory assays towards these enzymes were used for the investigation of hypocholesterolemic and antihypertensive compounds from pomegranate peel. The direct extract and its protein isolate were digested using two different enzymes followed by the bioassay evaluation of these hydrolysates. The obtained results were compared with that of the original extract and protein isolate. The peptides and polyphenols responsible for the bioactivity were identified by RP-HPLC-ESI-Q-ToF [50].

Multipotent compounds were isolated *in situ* from the leaves of *Helianthus annuus* L. by different bioassays in/on the HPTLC adsorbent bed. Two diterpenes-(-)-kaur-16-en-19-oic acid and 15-α-angeloyloxy-ent-kaur-16-en-19-oic acid, showing potent antioxidant, antibacterial, and cholinesterase inhibiting activities were isolated and characterized by a hyphenation of NMR-HRMS-DART-MS/MS [51]. Another combination of platelet aggregation and antithrombus assays together evaluated the antithrombotic properties of hawthorn leaves after fractionation. Twenty-five active compounds were identified from the active fraction by using

HPLC-QToF-MS technique, suggesting the presence of potent platelet aggregation inhibitors comprising monoterpenoids, diterpenoids, and flavones. These findings were supported by molecular modeling [52]. Turkish folk medicine for hemorrhoids was studied for the scientific evaluation of the therapeutic practice. *Verbascum lasianthum* Boiss. ex Bentham flowers, on activity-guided fractionation, identified to contain eight active compounds inserting antinociceptive and anti-inflammatory activities. The results were seconded by *in vivo* and toxicity studies [53].

Cold-pressed seed oils of flax, canola, and hemp are commercialized as premium products for their claimed health advantages. Compounds behind these properties were studied with the help of HPTLC, coupled with DPPH scavenging assay, AChE inhibition assay, pYES bioassay and antimicrobial *A. fischeri* bioassay and *B. subtilis* bioassays. The bioactive compounds separated by the target guides were analyzed by HPTLC-ESI-MS and some initial assumptions could have made [54].

## 5.9 BIOASSAYS FOR QUALITY CONTROL (QC)

Sometimes bioassays are placed in analyzes not only for preparation purposes, but as a necessary confirmation tool also. Many factors such as geographic origin, environmental, and growth conditions, agricultural practices and manufacturing processes contribute to the heterogeneous nature of natural formulations, which make the quality assurance of the herbal products a challenge, in terms of safety and efficacy. While using as a QC aid, the bioassay part evaluates and ensures the effectiveness, potency, and activity of the analyte. Quality bioassays can use any biomarker such as metabolites, enzymes, genes, or protein expression profiling, which can ascertain specificity and sensitivity of the testing analyte.

Safflower injection (Honghua injection), prepared from *Carthamus tinctorius* L. is a widely used TCM for many diseases, but is "black-marked" many times for its quality fluctuations. Keeping this in mind, a combination of chemical fingerprinting (CF), cell-based biological profile assay, and enzymatic assay was assigned to detect the quality deviations and its responsible components. CF by SYNAPT G2S-masslynx and data processing by UNIFI®scientific information system identified 33 compounds in the samples. The integration of CF and the two bioassays could identify all of the abnormal samples and the system furnishes a potential QC tool for other herbal formulations as well [55].

An interesting example for employing bioprofiling for the quality assurance is the quantitative method developed for bioprofiling of bioactive components from ginger (*Zingiber officinale* Roscoe) and ginger containing food products. Multipotent compounds with influence towards Radical scavenging activity, *A. fischeri*, *B. subtilis*, *pYES* (planar yeast estrogen screen), *AChE,* and tyrosinase inhibition were isolated from the ginger extracts and products. Along with the quantification of bioactive compounds-[6]-gingerol and [6]-shogaol, the procedure allows the identification of other multivalent bioactive components and measurement of bioactivity pattern and thereby ensures the product quality. In the HPTLC-HRMS hyphenation, the zones of interest were directly eluted into the HRMS, permitting the analysis of products containing even trace amounts of active components, enabling the characterization of all the potent compounds [56].

Another multiple biomarker assay was developed and checked the activity and quality of botanicals in 2017. They have tooled enzyme inhibition powered 'dual channel microfluidic chip' as the strategy for quality assessment and therapeutic consistency evaluation. One channel was designed for the enzymatic complex formation, where the other one dealt with the enzymatic reaction. A Chinese medicine-QiShenYiQi Pills (QSYQ), which consists of extracts of four herbs, was dissected by using thrombin and ACE. Eleven compounds were found to be inhibiting thrombin as well as ACE, and they were identified by their mass spectra. Drug quality assessment as well as screening of active compounds was achieved by the platform [57].

## 5.10 PITFALLS

While dealing with biosystems, many points are to be remembered. The instruments and processes should be necessarily clean and tidy, especially while repeating uses are there. When connected online, it should confirm that the bioassay media is compatible with the analysis system. Similarly, any pre-processing or conditions such as temperature, pH, etc., should not negatively affect the bioassay. The steps of the whole procedure, connected in series, should be adjusted to follow a moderate rate, to finish the sequences accordingly. In addition, many of the bioassays found to be a failure while evaluating the effects in terms of true biological effect and toxicity *in vitro*, being inferior to the results obtained from *in vivo* experiments. In addition, there are compounds showing their activity non-specifically towards bioassays, which are frequently appearing as false positives even in multiple bioassay screening and are named as pan-assay interference compounds (PAINS). The masking effect of abundant compounds on active minors is another factor, which can mislead the bioassay results.

## 5.11  CONCLUSION

Biological assays serve as the guiding methods for the characterization of biologically active compounds. All the methods follow a key feature where, a target-ligand interaction takes place in the provided biological system and the compounds which possess activity towards the specific function separate out from the extract, making further analysis easier. They can be organized well based on their execution and application. Keeping the view loyal to the topic, the presented chapter discussed only about bioassays compatible/used with HRMS for phytochemicals analyzes. Inside this dimension itself, there lie many opportunities for their utilization, such as natural product screening, metabolites stud, and preparation of libraries.

## KEYWORDS

- acetylcholinesterase
- angiotensin converting enzyme
- anticancer assays
- bioassay
- chemical fingerprinting
- combination bioassays
- planar yeast estrogen screen

# REFERENCES

1. Choudhary, M. I., & Thomsen, W. J., (2001). *Bioassay Techniques for Drug Development.* CRC Press.
2. Liu, Q., Li, S., Quan, H., & Li, J., (2014). Vitamin B12 status in metformin treated patients: Systematic review. *PLoS One, 9*(6), e100379.
3. Derosa, G., & Maffioli, P., (2012). α-glucosidase inhibitors and their use in clinical practice, *Arch. Med. Sci., 8*(5), 899–906.
4. Rother, K. I., (2007). Diabetes treatment-bridging the divide. *N. Engl. J. Med., 356*(15), 1499–1501.
5. Fujisawa, T., Ikegami, H., Inoue, K., Kawabata, Y., & Ogihara, T., (2005). Effect of two alpha-glucosidase inhibitors, voglibose and acarbose, on postprandial hyperglycemia correlates with subjective abdominal symptoms. *Metab. Clin. Exp., 54*(3), 387–390.
6. Schmidt, J. S., Lauridsen, M. B., Dragsted, L. O., Nielsen, J., & Staerk, D., (2012). Development of a bioassay-coupled HPLC-SPE-ttNMR platform for identification of α-glucosidase inhibitors in apple peel (*Malus domestica* Borkh.). *Food Chem., 135*(3), 1692–1699.S.
7. Li, Z. H., Ai, N., Lawrence, X. Y., Qian, Z. Z., & Cheng, Y. Y., (2017). A multiple biomarker assay for quality assessment of botanical drugs using a versatile microfluidic chip. *Sci. Rep., 7*(1), 1–11.
8. Guo, F., Zhang, S., Yan, X., Dan, Y., Wang, J., Zhao, Y., & Yu, Z., (2019). Bioassay-guided isolation of antioxidant and α-glucosidase inhibitory constituents from stem of *Vigna angularis. Bioorg. Chem., 87,* 312–320.
9. Murugesu, S., Ibrahim, Z., Ahmed, Q. U., Uzir, B. F., Yusoff, N. I. N., Perumal, V., Abas, F., et al., (2019). Identification of α-glucosidase inhibitors from *Clinacanthus nutans* leaf extract using liquid chromatography-mass spectrometry-based metabolomics and protein-ligand interaction with molecular docking. *J. Pharm. Anal., 9*(2), 91–99.
10. Silva, E. L. E., Lobo, J. F. R., Vinther, J. M., Borges, R. M., & Staerk, D., (2016). High-resolution α-glucosidase inhibition profiling combined with HPLC-HRMS-SPE-NMR for Identification of antidiabetic compounds in *Eremanthus crotonoides* (Asteraceae). *Molecules, 21*(6), 782.
11. Li, H., Song, F., Xing, J., Tsao, R., Liu, Z., & Liu, S., (2009). Screening and structural characterization of α-glucosidase inhibitors from hawthorn leaf flavonoids extract by ultrafiltration LC-DAD-MSn and SORI-CID FTICR MS. *J Am Soc Mass Spectrom., 20*(8), 1496–1503.
12. Ma, L. F., Yan, J. J., Lang, H. Y., Jin, L. C., Qiu, F. J., Wang, Y. J., Xi, Z. F., et al., (2019). Bioassay-guided isolation of lanostane-type triterpenoids as α-glucosidase inhibitors from *Ganoderma hainanense. Phytochem. Lett., 29,* 154–159.
13. Wubshet, S. G., Schmidt, J. S., Wiese, S., & Staerk, D., (2013). High-resolution screening combined with HPLC-HRMS-SPE-NMR for identification of potential health-promoting constituents in sea aster and sea rocket-new Nordic food ingredients. *J. Agric. Food Chem., 61*(36), 8616–8623.

14. Liu, B., Kongstad, K. T., Wiese, S., Jäger, A. K., & Staerk, D., (2016). Edible seaweed as future functional food: Identification of α-glucosidase inhibitors by combined use of high-resolution α-glucosidase inhibition profiling and HPLC-HRMS-SPE-NMR. *Food Chem., 203*, 16–22.

15. Wubshet, S. G., Moresco, H. H., Tahtah, Y., Brighente, I. M., & Staerk, D., (2015). High-resolution bioactivity profiling combined with HPLC-HRMS-SPE-NMR: α-glucosidase inhibitors and acetylated ellagic acid rhamnosides from *Myrcia palustris* DC. (Myrtaceae). *Phytochemistry, 116*, 246–252.

16. Wubshet, S. G., Liu, B., Kongstad, K. T., Böcker, U., Petersen, M. J., Li, T., Wang, J., & Staerk, D., (2019). Combined magnetic ligand fishing and high-resolution inhibition profiling for identification of α-glucosidase inhibitory ligands: A new screening approach based on complementary inhibition and affinity profiles. *Talanta, 200*, 279–287.

17. Wubshet, S. G., Tahtah, Y., Heskes, A. M., Kongstad, K. T., Pateraki, I., Hamberger, B., Møller, B. L., & Staerk, D., (2016). Identification of PTP1B and α-glucosidase inhibitory serrulatanes from *Eremophila* spp. by combined use of dual high-resolution PTP1B and α-glucosidase inhibition profiling and HPLC-HRMS-SPE-NMR. *J. Nat. Prod., 79*(4), 1063–1072.

18. Tahtah,Y., Wubshet, S. G., Kongstad, K. T., Heskes, A. M., Pateraki, I., Møller, B. L., Jäger, A. K., & Staerk, D., (2016). High-resolution PTP1B inhibition profiling combined with high-performance liquid chromatography-high-resolution mass spectrometry-solid-phase extraction-nuclear magnetic resonance spectroscopy: Proof-of-concept and antidiabetic constituents in crude extract of *Eremophila Lucida*. *Fitoterapia, 110*, 52–58.

19. Zhao, Y., Chen, M. X., Kongstad, K. T., Jager, A. K., & Staerk, D., (2017). Potential of *Polygonum cuspidatum* root as an antidiabetic food: Dual high-resolution α-glucosidase and PTP1B inhibition profiling combined with HPLC-HRMS and NMR for identification of antidiabetic constituents. *J. Agric. Food Chem., 65*(22), 4421–4427.

20. Zhao, Y., Kjaerulff, L., Kongstad, K. T., Heskes, A. M., Møller, B. L., & Staerk, D., (2019). 2 (5H)-furanone sesquiterpenes from *Eremophilabignoniiflora*: High-resolution inhibition profiling and PTP1B inhibitory activity. *Phytochemistry, 166*,112054.

21. Cádiz-Gurrea, M. D. L. L., Olivares-Vicente, M., Herranz-López, M., Arraez-Roman, D., Fernández-Arroyo, S., Micol, V., & Segura-Carretero, A., (2018). Bioassay-guided purification of *Lippiacitriodora* polyphenols with AMPK modulatory activity. *J. Funct. Foods, 46*, 514–520.

22. Okutan,L., Kongstad, K. T., Jäger,A. K., & Staerk, D., (2014). High-resolution α-amylase assay combined with high-performance liquid chromatography-solid-phase extraction-nuclear magnetic resonance spectroscopy for expedited identification of α-amylase inhibitors: Proof of concept and α-amylase inhibitor in cinnamon. *J. Agric. Food Chem., 62*(47), 11465–11471.

23. Zhao, Y., Kongstad, K. T., Jäger, A. K., Nielsen, J., & Staerk, D., (2018). Quadruple high-resolution α-glucosidase/α-amylase/PTP1B/radical scavenging profiling combined with high-performance liquid chromatography-high-resolution mass spectrometry-solid-phase extraction-nuclear magnetic resonance spectroscopy for

identification of antidiabetic constituents in crude root bark of *Morus alba* L.*J. Chromatogr. A, 1556*, 55–63.

24. Abbas-Mohammadi, M., Farimani, M. M., Salehi, P., Ebrahimi, S. N., Sonboli, A., Kelso, C., & Skropeta, D., (2018). Acetylcholinesterase-inhibitory activity of Iranian plants: Combined HPLC/bioassay-guided fractionation, molecular networking, and docking strategies for the dereplication of active compounds. *J. Pharm. Biomed. Anal., 158*, 471–479.

25. Schiffer, L., Arlt, W., & Storbeck, K. H., (2018). Intracrine androgen biosynthesis, metabolism, and action revisited. *Mol. Cell Endocrinol., 465*, 4–26.

26. Peters, R. J. B., Rijk, J. C. W., Bovee, T. F. H., Nijrolder, A. W. J. M., Lommen, A., & Nielen, M. W. F., (2010). Identification of anabolic steroids and derivatives using bioassay-guided fractionation, UHPLC/TOFMS analysis, and accurate mass database searching. *Anal. Chim. Acta., 664*(1), 77–88.

27. Shaari, K., Suppaiah, V., Wai, L. K., Stanslas, J., Tejo, B. A., Israf, D. A., Abas, F., et al., (2011). Bioassay-guided identification of an anti-inflammatory prenylated acylphloroglucinol from *Melicopeptelefolia* and molecular insights into its interaction with 5-lipoxygenase. *Bioorg. Med. Chem., 19*(21), 6340–6347.

28. Amorati, R., & Valgimigli, L., (2015). Advantages and limitations of common testing methods for antioxidants. *Free Radic. Res., 49*(5), 633–649.

29. Yang, B. Y., Yin, X., Liu, Y., Ye, H. L., Zhang, M. L., Guan, W., & Kuang, H. X., (2019). Bioassay-guided isolation of lignanamides with potential anti-inflammatory effect from the roots of *Solanum melongena* L. *Phytochem. Lett., 30*, 160–164.

30. Daud, M. N. H., Wibowo, A., Abdullah, N., & Ahmad, R., (2018). Bioassay-guided fractionation of *Artocarpus heterophyllus* L. J33 variety fruit waste extract and identification of its antioxidant constituents by TOF-LCMS. *Food Chem., 266*, 200–214.

31. Dash, U. C., Kanhar, S., Dixit, A., Dandapat, J., & Sahoo, A. K., (2019). Isolation, identification, and quantification of pentylcurcumene from *Geophilarepens*: A new class of cholinesterase inhibitor for Alzheimer's disease. *Bioorg. Chem., 88*, 102947.

32. Li, X. Y., Tang, H. J., Zhang, L., Yang, L., Li, P., & Chen, J., (2017). A selective knockout method for discovery of minor active components from plant extracts: Feasibility and challenges as illustrated by an application to *Salvia miltiorrhiza*. *J. Chromatogr. B, 1068*, 253–260.

33. Liu, Y., Nielsen, M., Staerk, D., & Jäger, A. K., (2014). High-resolution bacterial growth inhibition profiling combined with HPLC-HRMS-SPE-NMR for identification of antibacterial constituents in Chinese plants used to treat snakebites. *J Ethnopharmacol., 155*(2), 1276–1283.

34. Liu, Y., Staerk, D., Nielsen, M. N., Nyberg, N., & Jäger, A. K., (2015). High-resolution hyaluronidase inhibition profiling combined with HPLC-HRMS-SPE-NMR for identification of anti-necrosis constituents in Chinese plants used to treat snakebite. *Phytochemistry, 119*, 62–69.

35. Cretu, G. C., & Morlock, G. E., (2014). Analysis of anthocyanins in powdered berry extracts by planar chromatography linked with bioassay and mass spectrometry. *Food Chem., 146*, 104–112.

36. Móricz, Á. M., Häbe, T. T., Ott, P. G., & Morlock, G. E., (2019). Comparison of high-performance thin-layer with over pressured layer chromatography combined

with direct bioautography and direct analysis in real time mass spectrometry for tansy root. *J. Chromatogr. A., 1603*, 355–360.

37. Liu, M., Kipanga, P., Mai, A. H., Dhondt, I., Braeckman, B. P., De Borggraeve, W., & Luyten, W., (2018). Bioassay-guided isolation of three anthelmintic compounds from *Warburgiaugandensis* Sprague subspecies ugandensis, and the mechanism of action of polygodial. *Int. J. Parasitol., 48*(11), 833–844.

38. Kongstad, K. T., Wubshet, S. G., Johannesen, A., Kjellerup, L., Winther, A. M. L., Jager, A. K., & Staerk, D., (2014). High-resolution screening combined with HPLC-HRMS-SPE-NMR for identification of fungal plasma membrane H$^+$-ATPase inhibitors from plants. *J. Agric. Food Chem., 62*, 5595–5602.

39. Kongstad, K. T., Wubshet, S. G., Kjellerup, L., Winther, A. M., & Staerk, D., (2015). Fungal plasma membrane H -ATPase inhibitory activity of o-hydroxy benzylated flavanones and chalcones from *Uvaria chamae* P. Beauv. *Fitoterapia., 105*, 102–106.

40. Iqbal, K., Iqbal, J., Staerk, D., & Kongstad, K. T., (2017). Characterization of antileishmanial compounds from *Lawsonia inermis* L. leaves using semi-high resolution antileishmanial profiling combined with HPLC-HRMS-SPE-NMR.*Front Pharmacol., 8*, 337.

41. Uhlig, S., Hussain, F., & Wisløff, H., (2014). Bioassay-guided fractionation of extracts from Easter lily (*Lilium longiflorum*) flowers reveals unprecedented structural variability of steroidal glycoalkaloids. *Toxicon., 92*, 42–49.

42. Chua, C.,Lib, T., Pedersenb, H. A., Kongstad, K. T., Yana, J., & Staerk, D., (2019). Antidiabetic constituents of dendrobium officinale as determined by high-resolution profiling of radical scavenging and α-glucosidase and α-amylase inhibition combined with HPLC-PDA-HRMS-SPE-NMR analysis. *Phytochem. Lett., 31*, 47–52.

43. Han, Q. B., Zhou, Y., Feng, C., Xu, G., Huang, S. X., Li, S. L., Qiao, C. F., et al., (2009). Bioassay guided discovery of apoptosis inducers from gamboge by high-speed counter-current chromatography and high-pressure liquid chromatography/electrospray ionization quadrupole time-of-flight mass spectrometry. *J. Chromatogr. B, 877*(4), 401–407.

44. Segun, P. A., Ogbole, O. O., Ismail, F. M., Nahar, L., Evans, A. R., Ajaiyeoba, E. O., & Sarker, S. D., (2019). Bioassay-guided isolation and structure elucidation of cytotoxic stilbenes and flavonols from the leaves of *Macaranga barteri. Fitoterapia., 134*, 151–157.

45. Sangthong, S., Weerapreeyakul, N., Lehtonen, M., Leppanen, J., & Rautio, J., (2017). High-accuracy mass spectrometry for identification of Sulphur-containing bioactive constituents and flavonoids in extracts of *Raphanus sativus* var. *Caudatus alef* (Thai rat-tailed radish). *J. Funct. Foods, 31*, 237–247.

46. Pocasap, P., Weerapreeyakul, N., & Barusrux, S., (2013). Cancer preventive effect of Thai rat-tailed radish (*Raphanus sativus* L. var. *Caudatus Alef*). *J. Funct. Foods, 5*, 1372–1381.

47. Song, Y. H., Uddin, Z., Jin, Y. M., Li, Z., Curtis-Long, M. J., Kim, K. D., Cho, J. K., & Park, K. H., (2017). Inhibition of protein tyrosine phosphatase (PTP1B) and α-glucosidase by geranylated flavonoids from *Paulownia tomentosa*. *J. Enzyme Inhib. Med. Chem., 32*(1), 1195–1202.

48. Liu, B., Kongstad, K. T., Qinglei, S., Nyberg, N. T., Jager, A. K., & Staerk, D., (2015). Dual High-resolution α-glucosidase and radical scavenging profiling combined with

HPLC-HRMS-SPE-NMR for identification of minor and major constituents directly from the crude extract of *Pueraria lobate*. *J. Nat. Prod., 78*(2), 294–300.

49. Tahtah, Y., Kongstad, K. T., Wubshet, S. G., Nyberg, N. T., Jønsson, L. H., Jäger, A. K. Qinglei, S., & Staerk, D., (2015). Triple aldose reductase/α-glucosidase/radical scavenging high-resolution profiling combined with high-performance liquid chromatography-high-resolution mass spectrometry-solid-phase extraction-nuclear magnetic resonance spectroscopy for identification of antidiabetic constituents in crude extract of radix scutellariae.*J. Chromatogr. A, 1408*, 125–132.

50. Hernández-Corroto, E., Marina, M. L., & García, M. C., (2019). Extraction and identification by high-resolution mass spectrometry of bioactive substances in different extracts obtained from pomegranate peel. *J. Chromatogr. A, 1594*, 82–92.

51. Móricz, Á. M., Ott, P. G., Yüce, I., Darcsi, A., Béni, S., & Morlock, G. E., (2018). Effect-directed analysis via hyphenated high-performance thin-layer chromatography for bioanalytical profiling of sunflower leaves. *J. Chromatogr. A, 1533*, 213–220.

52. Gao, P., Li, S., Liu, K., Sun, C., Song, S., & Li, L., (2019). Antiplatelet aggregation and antithrombotic benefits of terpenes and flavones from hawthorn leaf extract isolated using the activity-guided method. *Food Funct., 10*(2), 859–866.

53. Kupeli, E., Tatli, I. I., Akdemir, Z. S., & Yesilada, E., (2007). Bioassay-guided isolation of anti-inflammatory and antinociceptive glycoterpenoids from the flowers of *Verbascumlasianthum* Boiss. ex Bentham. *J. Ethnopharmacol., 110*(3), 444–450.

54. The, S. S., & Morlock, G. E., (2015). Analysis of bioactive components of oilseed cakes by high-performance thin-layer chromatography-(bio) assay combined with mass spectrometry. *Chromatography, 2*(1), 125–140.

55. Feng, W. W., Zhang, Y., Tang, J. F., Zhang, C. E., Dong, Q., Li, R. Y., Xiao, X. H., et al., (2018). Combination of chemical fingerprinting with bioassay, a preferable approach for quality control of safflower injection. *Analytica Chimica Acta, 1003*, 56–63.

56. Krüger, S., Bergin, A., & Morlock, G. E., (2018). Effect-directed analysis of ginger (*Zingiber officinale*) and its food products, and quantification of bioactive compounds via high-performance thin-layer chromatography and mass spectrometry. *Food Chem., 243*, 258–268.

57. Li, Z. H., Ai, N., Yu, L. X., Qian, Z. Z., & Cheng, Y. Y., (2017). A multiple biomarker assay for quality assessment of botanical drugs using a versatile microfluidic chip. *Sci. Rep., 7*, 12243.

# CHAPTER 6

# Bioanalytical Screening/Purification Techniques

SHINTU JUDE and SREERAJ GOPI

*Research and Development (R&D) Center, Plant Lipids (P) Ltd., Kadayiruppu, Kolenchery, Cochin, Ernakulam, Kerala – 682311, India*

## ABSTRACT

There are newer technologies that happen to arise even within days, regardless of the fields, application, related instruments, etc. While considering the phytochemicals, owing to the nature of the matrix, all the related processing steps such as the extraction, isolation, and analysis seem to be tedious and time-consuming. Hence, many screening techniques are developed for the feasible isolation of bioactive compounds. Herein the chapter, the leading technologies in bioanalytical screening methodologies including TLC bioautography, ultrafiltration, ligand fishing, HPLC based post-column bioassays, high performance displacement chromatography, affinity chromatography, cell membrane chromatography (CMC), size exclusion chromatography (SEC), etc., are discussed with relevant examples.

## 6.1 INTRODUCTION

The bioactivity-guided characterization and isolation often fail in the case of less abundant bioactive compounds. If the number of active compounds present in the matrix, the relative concentration differences also play a villain role in the bioactive characterization. Plant secondary metabolites constitute most of the bioactive components of nature. They represent ligands of biological targets, because of their structural and functional peculiarities. Taking advantage of this property, screening techniques based on functional responses, secondary activities or interactions between some targets are used in methodologies for the active compounds isolation and screening. Thus, the next stage of developments in the field of active components identification introduced some analytical techniques. In this chapter, we discuss about some leading analytical-screening techniques, which can be coupled with HRMS. A presentation of all the techniques is given in Table 6.1.

**TABLE 6.1**  The Leading Technologies for Bio-Analytical Screening

| Technique | Working Principles | References |
|---|---|---|
| TLC bioautograghy | Separation based on localizing the activity of complex matrix | [1–14] |
| Ultrafiltration | Membrane-based separation technique | [15–26] |
| Ligand fishing | Receptor-ligand affinity adsorption | [27–33] |
| HPLC based post-column assays | HPLC separation followed by a protein-ligand interference | [35–56] |
| Affinity chromatography | Ligand affinity separation | [60–63] |
| Cell membrane chromatography | Affinity separation utilizing cell membrane with specific receptors | [65–85] |
| Size exclusion Chromatography | Separation of Ligand bounded large compounds from the unbound small molecules | [86, 87] |

## 6.2  TLC BIOAUTOGRAGHY

Bioautography correlates chromatographic separation with bioactivity screening. TLC/ HPTLC serves as the chromatographic segment in many bioautography analyzes due to its rapid localization efficacy on the active compounds from complex matrices, and is hence considered here as the prime approach. A biological target prepared in a suitable medium is applied on a developed TLC plate, and placed for incubation. The separated zones proceeded for further analyzes. Earlier studies dealt only with antifungal compounds. Now, a number of biological targets such as enzymes, bacteria, yeasts, oxidation processes, etc., can be applicable to TLC bioautography. By using bioautography, the effective isolation of compounds from a complex matrix and an improved identification of their activity can be achieved. Therefore, bioautography is preferred for the investigation of natural products.

Depending on the working, bioautography can be classified into three, Contact bioautography, which allows a direct contact of bioactive compounds from the TLC plate by transferring to the inoculated medium. As a result, growth inhibition zones appear in the contact places, corresponding to the active compounds, second one is immersion method, in which the developed TLC plate is immersed (or covered with) in the medium. Medium remains on the surface throughout the incubation time and visualization. The third method is named as direct bioautography. Here, both the separation and bio screening are conducted on the chromatography plate directly. The microbes are grown directly on the developed TLC plate. In the case of antioxidant assays, though they are considered as bioassays, the actual procedure does not bear any biological processes, methods or aids, but only chemicals. Here, in one mode, the assay medium is kept in contact with the separated test compounds on TLC plate, is by spraying the same onto the plate.

The working of TLC bio-autography is presented in Figure 6.1. The active compounds are isolated thoroughly based on their bioactivity and the applied profiling procedure, irrespective of their nature, functional groups, or abundance. For example, alkaloids from five different *Amaryllidaceae* species were studied for their AChE inhibitory activities by using PLE (pressurized liquid extraction)/SPE/HPLC/ESI-octopole-oaToFMS hyphenation. Around 17 individual alkaloids were structurally identified from the complex mixture, including the newly determined dihydrogalanthamine [1]. In another study, meroterpenes were identified as the nitric oxide (NO) production inhibiting molecules from *Psoralea corylifolia* fruits, one of the well-utilized plants in traditional Chinese medicine (TCM). Isolated compounds-one new and two established-were structure elucidated with NMR-HRMS-X ray crystallography techniques, and their activity was further confirmed with Nitrite assay on LPS-induced NO production in RAW 264.7 cells [2]. Antibacterial compounds separated from *Eugenia jambolana,* were alkaloids, flavonoids, saponins, anthraquinones, glycosides, terpenoids, steroids, and phenolic compounds [3]. One earlier attempt to visualize the interaction of lipase with the inhibitory components in the lotus (*Nelumbo nucifera* Gaertn.) leaves have successfully recorded six alkaloids as the active components from their $MS^n$ (n = 4) data by coupling TLC bioautography with electrostatic field induced spray ionization [4].

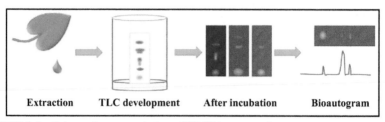

| Extraction | TLC development | After incubation | Bioautogram |

**FIGURE 6.1**　Schematic representation of TLC bioautography.

Antimicrobial investigation is a very interesting domain handled with TLC bioautography. TLC bioautography was utilized for the screening of compounds having antifungal activity against *Candida albicans* from the stem bark extract of *Croton heliotropiifolius* Kunth. The study made use of HPLC-ToF-HRMS for the identification and structure elucidation of the nine isolated compounds [5]. In the same way, bioactive compounds of *Tanacetum vulgare* L. were investigated with HPTLCUV/Vis/FLD-EDA-HRMS. The bioautography was accomplished by four antibacterial cultures and the compounds in the bioactive zones were identified as polyacetylenes, with an orbitrap MS [6]. Along with the usual TLC-bioassay-HRMS pattern, Jamshidi-Aidji et al. used inverse densitometric measurement for the quantification of the inhibition zones. In addition, they have given a statistical demonstration for the effect of different media on the results [7].

TLC, in hyphenation with more than one bioactivity profiling enables the recognition of multipotent compounds. By using a combination of TLC-DPPH-Hydroxyl radical scavenging assay-Erythrocyte membrane stabilization assay, an antioxidant component having erythrocyte membrane stabilizing potential was identified from *Carissa carandas* (L.) leaves extract. The following TLC-HPLC-UV-FTIR-GC-HRMS setup was used for the characterization of the separated active compound as 20-hydroxypregnan 18-oic acid [8]. Both antifungal and antibacterial activity were shown by *Combretum molle* leaves, as proved by TLC followed by inoculation with a spray of active suspension of microbial cells and the strong inhibition zones were found to be corresponding to the active compounds [9]. In the same way, the active compounds were screened by the HPTLC-antibacterial, acetylcholinesterase (AChE) screening-HRMS profiling from *Salvia miltiorrhiza* Bunge root [10]. A potential antioxidant, with strong cholinesterase inhibitor properties was isolated from *Geophila repens* (L.) I. M. Johnst (Rubiaceae), by using a combination of bioassays and TLC bioautography [11].

HPTLC was implemented as a quantitative aid and as a bioanalytical screening tool. These two dimensions were demonstrated together in the discovery of bioactives from *Geophila repens*. An antioxidant, effective for different oxidants as well as oxidative damage of the cells was isolated and was recognized as Pentylcurcumene, by applying NMR and HRMS. The quantification attempts of the study resulted in a sensitive and precise quantification of compound accurate in nanogram levels. The active compound was further demonstrated to possess inhibitory activity against AChE and BChE. Mode of inhibition together with molecular docking details and enzyme kinetics data suggest the application of Pentylcurc-umene even in high-end therapeutics, such as AD treatment [12].

The possibility of TLC bioautography to handle different samples simultaneously, makes the analyzes and comparisons of phytochemicals rapidly and hence the bioactive compounds against the requirement can be selected easily. From a group of ten different Mediterranean plants studied against seven different bacteria strains, *Diplotaxis harra* exhibited the best antibacterial activity. The separation, identification, and structural characterization were carried out and the invention of active compound Sulforaphane, holding a strong history of therapeutic benefits in its name, was seconded by the confirmed antibacterial activities [13].

Many shortcomings are reported for TLC bioautography so far. As the process involves two steps, the number of influencing factors is also high. The chromatographic elements such as solvents, buffers, sample preparation agents, etc., can affect the second step involving microbial processes. pH and temperature of the fluids from chromatographic steps may also interfere with the microbes, especially, the acidic conditions have a destroying effect on the bioassay response [14]. In addition, the range of enzymes and other microbial are restricted as the amount needed for detection is high in many cases and it is not practical for all the microbes to obtain in that much higher amounts. Besides, thorough knowledge on the microbes and procedure outputs are needed for the appropriate selection of test microorganism and incubation. Again, if the active compounds exhibit multivalent actions, a single assay may not be enough for the differentiation between the activities; a second or third assay is recommended in these cases. Then, in addition, many of the TLC bioautographies constrain its usage for quantitative analysis. Moreover, many biological assays are not compatible with the usual TLC conditions.

## 6.3 ULTRAFILTRATION

Ultrafiltration is a membrane-based technology, where the separation depends on the molecular size. It was originally developed for the screening of combinatorial libraries, and was then used for the processing of macromolecules and finally for the testing of bioactive compounds. As seen in Figure 6.2, the extract of analyte allows incubating with the target protein, at specific temperature, for a particular time period. Then ultrafiltration is performed by using ultrafiltration membrane having desired molecular weight cut off and washed in order to remove the unbound matrix. The ligand-protein complexes are chemically treated to dissociate the bond-in-between. It is used for the structural investigation of natural products as a preparative tool along with mass instruments. Ultrafiltration can be of two modes-offline and online. In offline ultrafiltration, the filtrates were preceded for further analysis manually. In online procedure, the analysis setup will be coupled to the ultrafiltration chamber automatically and is known as pulsed ultrafiltration [15].

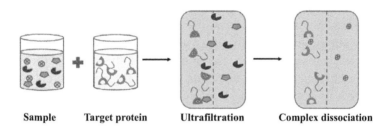

Sample          Target protein          Ultrafiltration          Complex dissociation

**FIGURE 6.2**   Working of ultrafiltration.

Offline combination of ultrafiltration-LCMS was successfully used for the investigation of quinone reductase-2 inhibitors from *Humulus lupulus* L. which is important for its antimalarial, antitumor activities. The ligands were identified and structure elucidation was carried out [16]. A different study handled the investigation of α-glucosidase inhibitory compounds from the leaf flavonoid extract of a traditional medicinal plant, *Crataegus oxyacantha* L. commonly known as Hawthorn by α-glucosidase inhibition assay, followed by the ultrafiltration LC-DAD-MS[n]. The α-glucosidase ligands separated by ultrafiltration were identified by FTICRMS [17]. HPLC-DAD-MS[n] worked well with ultrafiltration in the profiling of tyrosinase inhibitory compounds from the mulberry leaves extract (*Morus alba*) [18]. The tyrosinase inhibitors were isolated and characterized from 212 metabolites of *Gastrodia elata* using an offline combination of ultra-filtration-UV-MS-NMR [19]. Cyclooxigenase-2 inhibitors were another vital herbal compound isolated by using affinity ultrafiltration from Zi-shen pill, a well-known Chinese medicine for benign prostatic hyperplasia. By using HRMS, eight active compounds were identified, and COX-2 inhibition assay as well as molecular docking studies provided the supporting molecular level confirmation on the functioning of active molecules [20]. Further, a study from Song et al. developed an ultrafiltration-based strategy to remove unspecific binding of ligands by adding another ligand with more affinity and thereby inserting an "enzyme channel blocking." Here, during the screening of Flos Chrysanthemi for xanthine oxidase (XOD) inhibitors, febuxostat was made bind to the channel, in order to prevent binding of ligands. Therefore, the four identified XOD inhibitors were believed to be much stronger than the blocking compound febuxostat, which was further verified by microplate inhibition assays [21].

Pulsed ultrafiltration is an online; MS-based technology, where a pulse of ligands is injected to a cell containing macromolecular receptors enclosed by ultrafiltration membrane. The binding between ligands and macromolecules alters the elution profile. The ligand-macromolecular complexes, after washing, are disrupted by appropriate simple preparation methods and the ligands to be analyzed are directed to MS. As pulsed ultrafiltration provides more convenience and advantages over conventional ultrafiltration in terms of amount of analyte and reusability of targets, it finds a better place in applications [22]. Investigation of estrogenic activities of eight different herbal preparations, conducted by using a bioassay guided pulsed ultrafiltration-LC-MS is an excellent example for the application of pulsed ultrafiltration [23]. The selectivity and efficiency of using pulsed ultrafiltration were validated by the selective screening of anti-inflammatory compounds from 18 different matrices enriched with one or two COX-2 inhibitors. Mass spectral data verified the presence of active compounds, and the lack of interference from the sample matrix, irrespective of nature [24]. The presence of selective and non-selective inhibitors of COX-1 and COX-2 was confirmed in eleven botanicals used for the TCM, Huo-Luo-Xiao-Ling Dan. The seventeen identified active compounds were studied for their inhibiting activity towards human COX-1 and COX-2 [25].

Despite the potential effects and applications of ultrafiltration-both in offline and online modes, it has some limitations. The ease of use is restricted by the fact that; it doesn't suit transmembrane proteins which may cause non-specific interactions. Unavailability of respective proteins in sufficient amount is another factor which limits the strategies of investigations. In addition, in most of the cases, the compounds having more activities and high abundance shows more affinity towards the protein binding, and the method overlooks the low affinity, though synergy-contributing ligands [26].

## 6.4    LIGAND FISHING

Ligand fishing is a technique based on receptor-ligand affinity adsorption, where the receptors are immobilized on a support and this receptors are used for 'fishing out' the ligands/active compounds present in the complex matrix, which have an affinity towards the receptor, and the non-binding molecules will be remained in the matrix. The active compounds co-separated with the receptors can be eluted out for further analysis [27, 28]. Normally, magnetic nanoparticles are used as the solid support and proteins are as the receptors. Even the less abundant metabolites could be identified by this technique, if they possess affinity towards the ligand protein. Figure 6.3 can narrate it more clearly.

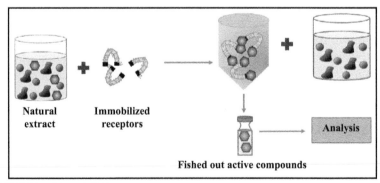

**FIGURE 6.3**    Ligand fishing.

In the primary stage of development of "ligand fishing" processes, many of the studies were conducted on alpha glucosidase (AGH) inhibitory compounds. In such a study, N-terminus-coupled AGH magnetic nanoparticle beads (AGN-TCMB) were prepared and their ligand fishing properties were identified by Wubshet et al. They have successfully characterized the crude extract of *Eugenia catharinae* O. Berg. as a constitution of AGH inhibitory ligands, alkyl resorcinol glycosides and flavonoids by magnetic ligand fishing-HPLC-HRMS-SPE-NMR hyphenation [29]. Deng et al. introduced a modified version of this layout, in which the "control support" without AGH was also included for the screening, along with the AGH immobilized on CNBr activated sepharose beads, so that the false positives could be eliminated by comparing both. The effectiveness of the method was demonstrated by fishing out three AGH inhibitors from the crude extract of green tea, which were identified as epigallocatechin gallate (EGCG), gallocatechin gallate (GCG) and epicatechin gallate (ECG), by analyzing with QToF-MS [30].

Ligand fishing-HRMS coupling provided a new face for antidiabetic therapeutic herbals in many studies. Utilization of agonists of peroxisome proliferator-activated receptor-γ (PPARγ) was catalyzed by their scope of application as antidiabetic agents, and one of the important study among them tooled fusion protein affinity chromatography. Human PPARγ ligand binding domain in fusion with glutathione S-transferase (GST-hPPARγLBD) was allowed for the selective attachment of GST with glutathione and was then used for the direct selective identification of interacting compounds from *Dendranthema indicum* flowers. The rapid, low cost, non-denaturing, renewable, high throughput technology, in combination with HPLC-ESI-Q-ToF-MS/MS identified the active compound as Isochlorogenic acid A [31].

Tao et al. demonstrated the scope of multi-target immobilization and its utilization in compound screening. Maltase, lipase, and invertase were selected as the targets on the magnetic beads for screening antidiabetic compounds from the complex Chinese traditional herbal medicine "Tang-Zhi-Qing." A detailed characterization including the morphology, ligand screening conditions, compound specificity, kinetics, etc., were verified for the prepared magnetic beads. Further, five active compounds were identified and their activities were validated, including the synergic activities [32].

Ligand fishing was introduced with the advantages of identifying even the compounds of low affinity and the high affinity compounds of less abundance. Thrombin was such a compound, which is known as the principle enzyme of hemostasis and take part in many important functions in the body, including the usage as diagnosis biomarker, blood clotting enhancement, etc. *Salvia miltiorrhiza* Bunge roots, commonly known as

DanShen was traditionally used for the treatment of platelet coagulation, thrombosis, etc., and so, its extract was of high interest in characterization studies. Thrombin, covalently bonded with magnetic beads were incubated along with DanShen injection and the bound compounds were characterized. Two of the active compounds-protocatechuic aldehyde and salvianolic acid C-were further evaluated with a traditional inhibitory assay. The conclusion of the study suggests the association of inhibitory activities of the compounds with their specific structures [33].

## 6.5   HPLC-BASED POST-COLUMN ASSAYS

In the basic form, post-column bioassays consist of an HPLC separation followed by interaction of compound of interest with target protein, which then reacts with the reporter ligand. The reaction products are analyzed through UV, fluorescence or MS detectors, and the presence of an inhibitor gives a negative peak, as it prevents the reaction product formation. The process is mentioned as online-bioassay profiling (Figure 6.4(a)). In some cases, the eluate from column is run parallel for dereplication. In one of the first applications of this online bioassay for phytochemical bioactives, AChE inhibitors were identified from a natural matrix by coupling HPLC-UV-MS with biochemical detection using Ellman's reagent [34]. Later, the technology was presented by Kchaou et al. with slight modifications for the identification of AChE inhibitors from *Zygophyllum album* extracts [35]. The further developments in this technology have replaced UV from the series, first with fluorescence readouts, then with MS readouts [36, 37].

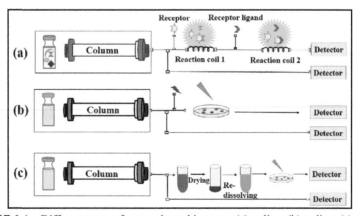

**FIGURE 6.4**   Different types of post-column bioassays (a) online; (b) at-line; (c) offline.

In post-column assays, determination of antioxidant activity employs different assays such as ROS, DPPH, ABTS, etc., for the detection [38]. A post-column online DPPH assay, together with NMR and QToF HRMS detailed the antioxidants from *L. tibetica* extract as pinoresinol-β-D-glucoside, isoacteoside A, acteoside, tibeticoside, epipinoresinol, anthelminthicol A and phillygenol [39]. Inhibition of XOD is important in treatment of gout and many other related diseases. A post-column dual activity assay-XO inhibition and antioxidant assays-distinguished the bioactive compounds from *Oroxylum indicum* extract, which were identified by UV and mass spectral data [40]. In another study, a complete phytochemical characterization was conducted for *Schisandra chinensis* s (Turcz.) Baill. using HPLC-ESI-ToF-MS. Out of the characterized compounds, only one liganan-Gomisin D, along with quercetin glycosides and the chlorogenic acid isomers were potent enough to inhibit ABTS+ radicals by TEAC measurements [41]. Again, ABTS assay, coupled parallel with MS was able to identify the antioxidants from *Angelica sinensis* essential oil [42]. At the same time, antioxidant capacity of pigments from red cabbage, perilla, and elderberry were evaluated by DPPH radical scavenging assay and the compounds behind the activity was pinpointed by NMR and HRMS [43]. The same instrumentation was used to determine the real particles behind the potential of Yangxinshi tablet (YXST) in the treatment of cardiovascular diseases. By aiding HPLC-ESI-Q-ToF-MS, a total of 127 compounds from different category were identified, and were assessed for their antioxidant activity *in vivo* and *in vitro* [44].

Nine bioactive compounds, inhibiting the growth of *Plasmodium falciparum* were isolated by microfractination with HPLC from *Carica papaya* leaf extracts, which have proven to be antiplasmoidal *in vivo* and *in vitro*. A complete activity profiling consisting of *in vitro* cytotoxicity assay and activity against *Plasmodium falciparum* and *Trypanosoma brucei rhodesiense* together with *in vivo* activity against *Plasmodium berghei* was conducted and the results suggest piperidine alkaloids as the active compounds *in vivo,* though lacking their activities *in vitro*. These findings were further affirmed with the structure elucidation and differentiation of the compounds by HRMS and NMR [45]. In another study, identification of three estrogenic compounds from pomegranate (*Punica granatum*) was fulfilled by online beta-estrogen receptor bioassay coupled to HPLC separation [46]. The possibilities of DNA binding were well used in screening

*Trollius chinensis* Bunge. The total analysis setup of HPLC-DAD-ESI-IT-ToF-MS-DNAEB-FLD resulted in the isolation of 16 and identification of six active compounds [47].

The capacity of multi-target bioassays for the target estimation of active compounds in a single chromatographic run was well utilized by Li et al. for the investigation of XOD inhibitory and free radical scavenging components from *Oroxylum indicum* extract. Interestingly, the method detected seven potent compounds, including a dual active compound [48].

Different assays intended for the investigation same category of compounds may provide data with different manners. In a study of anti-oxidants, oxygen radical absorbance capacity (ORAC) assay connecting online with HPLC provided sensitive results for radical scavengers, while DPPH assay gave more selective results, even for rapid radical scavengers. Compounds without authentic standards also were identified by HRMS, and confirmed with NMR from *Cyclopia genistoides*, while Iriflophenone-3-C-glucoside, isomangiferin, and mangiferin, being the major active compounds [49].

A better resolution was obtained by slightly modifying the procedure, and named "atline bioassay profiling" (Figure 6.4(b)). The eluate from the column is fractionated and collected in microplates, (not proceeded to reaction coil) to carry out bioassay. This concept was successfully implemented in the simultaneous parallel analysis for the bioactivity profiling by enzyme assay and compound identification by Q-ToF MS data acquisition of *Cistus incanus* active extract [50].

Another modification causes the name "offline profiling" which was made to the procedure, where the post column eluate is collected, dried, and reconstituted with suitable solvent, then completed the bioactivity profiling in different microplates (Figure 6.4(c)). Offline profiling allows a coupling with other detection aids such as NMR also. Offline profiling was used for the bioactivity profiling of different extracts of *Isatis tinctoria* L. against cyclooxygenase (COX)-2 inhibitory activity and identified tryptanthrin (1) as the bioactive component [51]. A simple procedure for the dereplication of plant extracts containing GABA(A) ligands was demonstrated with *Xenopus oocytes* by using a gradient elution fractionation connected with a two-microelectrode voltage clamp assay [52]. The post-column bioactivity assays have used to assess plant extracts for their antiprotozoal, anti-trypanosomal, antiplasmodial, and antiretroviral activities and DYRK1A kinase inhibitory activity [38]. By incorporating NMR in the instrumentation setup, the full capability of dereplication analysis setup was achieved and the platform was demonstrated by the identification of benzophenanthridine alkaloids in *Eschscholzia californica* cell culture [53]. By combining semi-preparative HPLC-MS fractionation along with capillary NMR and bioactivity assessment, it was a successful platform for the identification of bioactive components from *Rhynchosia viscose* [54]. Jonker et al. demonstrated the utilization of GC as a preparative technique and have successfully tooled with the AR-EcoScreen reporter gene bioassay [55].

However, in contradiction, sometimes the lack of factors such as proper selection of reaction patterns, strong activity, sufficient amount of compound for the reaction, etc., causes the active molecules not to appear in the assayed matrix while comparing with the other activity measurements. In addition, in natural extracts, the assayed results may be obtained due to the synergic effects of compounds with weak activities or trace amounts only, not from any single potent molecule. For example, *Curcuma wenyujin* essential oil didn't show the presence of any corresponding antioxidants in HPL-ABTS assay, regardless its verified *in vitro* antioxidant activities [56]. So, in such cases, it is preferred to enrich the samples prior to the assays.

## 6.6 HIGH PERFORMANCE DISPLACEMENT CHROMATOGRAPHY

Displacement chromatography is primarily a preparative purification technique, and has been applied in many studies for the purification of biological/ biochemical entities. Displacement chromatography works on the fact that; only finite number of binding sites will be available on the surface of the stationary phase. Therefore, if one site is occupied by a molecule, then it will be unavailable for the others. Hence, in equilibrium, the molecules with high affinity towards the stationary phase will bind more, leaving the lesser affinity particles with mobile phase flow. A suitable displacer displaces the adsorbed sample components from the stationary phase and as result, occupation of binding sites proceeds with the order of affinity. The continuous displacements finally result in consecutive rectangular zones of pure compounds. Application of displacement chromatography as preparative purification method was demonstrated by the studies on *Enantia chlorantha* Olive. The bioactive alkaloids were isolated from the stem bark extract and were identified as palmatine and dl-tetrahydropalmatine by HRMS and NMR [57].

## 6.7    AFFINITY CHROMATOGRAPHY (BIO-AFFINITY CHROMATOGRAPHY)

Affinity chromatography can be considered as a kind of liquid chromatography, where the stationary phase is an affinity ligand-a biological binding agent-which is immobilized in a column, in order to bind the active components selectively from the sample under specific conditions. After removing the impurities by washing, the ligands were collected by eluting with proper eluent. Better instrumentation setups were introduced for affinity chromatography, depending on the binding agent. In frontal affinity chromatography (FAC), continuous infusion of complex matrix containing the analyte molecule is provided to the stationary phase. The delay in elution till the active molecules come out of the column, after exceeding its binding capacity is compared with that of a reference compound and mentioned as the binding affinity of ligands. Figure 6.5 presents a clear picture on this.

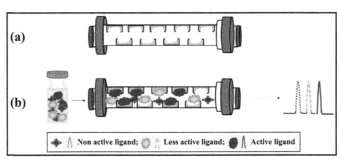

**FIGURE 6.5**    (a) A stationary phase in frontal affinity chromatography; and (b) its effect on the separation of a mixture of active, less active and non-active ligands.

The immobilized biomolecule target can be an isolated molecule such as enzyme, protein receptor, etc., transmembrane proteins or even live cells. While using enzymes as the biomolecule target, immobilized enzyme reactors (IMERs) provide a more convenient platform to analyze the bioactivity of the natural extract. Here, a substrate, specific to the target enzyme is inserted along with the extract, such that the substrate is catalyzed by the enzyme and produce specific products. However, the bioactive compounds present in the natural product extract bind with the enzyme and thereby prevent the formation of these products. This allows the measurement of inhibitory activity of compounds from a mixture towards specific enzymes (Figure 6.6) [58].

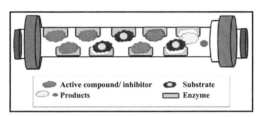

**FIGURE 6.6**   Immobilized enzyme reactors (IMERs).

The standard properties of affinity chromatography and HRMS perform together in a synergic way, while coupling, as well as target ligand interaction studies. In a study, antibodies were immobilized in the column to act as the stationary phase and were used for the characterization of the phenolic compounds in *Phyllantus urinaria* extract [59]. Nonetheless, a protein affinity chromatography was used to screen the PPAR$\gamma$ ligands from *Dendranthema indicum* flowers [60]. By immobilizing $\beta_2$ adrenergic receptor on silica, four $\beta_2$-AR-targeting compounds were separated from corydalis rhizomes. A coupling of the high-resolution mass spectrometer (MS) ToF enabled the system to identify the active compounds. Moreover, the interaction pattern, as well as the binding mechanism, was established to demonstrate the scope of utilization [61]. Eighty-seven putative 14-3-3 client proteins from rice (*Oryza sativa* L.cv.) roots were identified and their functional characterization was carried out with affinity chromatography-HRMS platform. The beauty of the investigation lied on the collected information on the isoform specific functions of 14-3-3 proteins in root including the growth, carbon metabolism, trafficking, self-defense mechanism, energy metabolism, amino acid metabolism, cell structure and development, signaling, binding, and transport [62].

A complete interchange in the application priority and functions between affinity chromatography and HRMS compound identification was observed in a study from Turkey. Here, 10 new "reverse inhibitors of peroxidase (RIPs)" were prepared synthetically and confirmed by NMR and HRMS. These RIP's were further used for a rapid, cost-effective purification of peroxidase enzymes from radish species with high yield [63].

Affinity chromatography gives up its points in two factors, one is the possibilities of false-positive results, which could occur from non-specific binding of compounds, and 2nd is the lack of provision for the simultaneous analysis of bounded compounds for their activities.

## 6.8   CELL MEMBRANE CHROMATOGRAPHY (CMC)

Cell membrane chromatography (CMC) is a kind of affinity chromatography, where the stationary phase is fabricated with the cell membranes, which contain specific receptors and are immobilized on a silica carrier. As it is possible to construct the cell lines with expressions of specific receptors, the stationary phases can be customized with definite properties such as controllable expressions, uniform stationary phase surface, improved method specificity, and selectivity as well as increased sensitivity for components [64]. Cellular membrane affinity chromatography was named after the columns with immobilized transmembrane proteins. To date, the technology was mostly reported to be used for the characterization of components in TCM herbs. Antiplatelet components screening from 'Danshen' using platelet CMC- UHPLC-QToF-MS/MS is an example for this [65]. *Corydlis Decumbentis Rhizoma* was investigated for its β1 adrenergic receptor (β1-AR) inhibitors by the same analytical configuration, differing only in the mounted cell membrane, here which is β1-AR. The bone diseases counteracting compound-jatrorrhizine was isolated and recognized by the platform [66].

Active compound and their metabolites could have screened even from the *in vivo* urine samples, post administration of *Aconitum carmichaeli* roots by a rat cardiac muscle cell CMC-ToF/MS. The identified 24 active alkaloids and 10 metabolites thereof demonstrated the structure-affinity relationship and retention behaviors, along with a possibility of semi quantification of compounds in fractionated samples [67]. The same instrumental layout was used to analyze the very same sample in another study-but the intention was different. Here, the behavior of the active components both on normal and pathological cells was investigated by using doxorubicin (DOX)-induced heart failure as the model. The analytical framework was considered to be two-dimensional (2D), as the fractions retained and recognized in the affinity column, which acts as the first dimension was introduced offline or online to the second dimension, mostly consisting of a monolithic column and HRMS [68]. Out of the 16 active compounds, not surprisingly, four components were proved to specifically counteract the heart failure, which was further confirmed by an *in vitro* pharmacodynamic examination of the most active one of them-talatizamine [69]. Another comparative 2D CMC have run for the identification of anti-hepatoma components from *Scutellariae Radix*. Simultaneous loadings on hepatic carcinoma HepG2-CMC columns and normal hepatic L02 CMC columns resulted in the distinction of three efficient, selective antitumor components, recognized as oroxylin A, wogonin, and chrysin by QTof. Cell proliferation and toxicity studies were in agreement with the recommendation of their usage as therapeutics [70]. By comprising hepatocarcinoma cell line SMMC-7721 in the same HepG2-CMC layout, more efficient and synergic anticancer components-Adenosine, and Bruceine B were isolated [71]. Another study has attracted interest for the rapid screening of 28 herbal medicines for epithelial cell growth factor receptor (EGFR) by exploiting 2D HepG2/CMC-ToFMS. The detected four active components were shown to be selective inhibitors of EGFR [68]. CMC incorporating human epidermal squamous cells ($A_{431}$) cells was employed in a study, targeting EGFR antagonists from medicinal herbs. The active compounds-oximatrine and matrine were identified from *Radix sophorae flavescentis* by mass patterns and were subjected to competitive displacement assay, *in vitro* EGFR secretion assay and MTT cell growth assay for assessing their functional properties [72].

Human embryonic kidney 293 cells (HEK293) are one of the most used cell lines for academic and research purposes owing to their properties, and so in CMC. One important CMC setup significant in the investigation of antitumor compounds is HEK293/VEGFR-CMC. The antitumor activity of *Corydalis Thizoma* was shown attributed to Tetrahydropalmatine and corydaline [73]. Again, the antitumor effect of *Rhizoma Belamcandae* was investigated by using HEK293/EGFR-CMC-HPLC-IT-ToF-MS [74]. HEK293 based CMC utilization is a real potential aid, which can be tailored as per the requirements, as shown in the screening of $\alpha_{1A}$ adrenoceptor agonists from *Peucedanum praeruptorum Dunn* utilizing HEK293 $\alpha_{1A}$/CMC and characterization of tumor antagonists from *Brassica albla* by using HEK293/FGFR4-CMC [75, 76].

A more complicated matrix of MaiLuoNing injection-a traditional Chinese therapeutic, containing five herbal extracts in it, is used in many treatments, but are mostly associated with many anaphylactic reactions.

So, by using Rat basophilic leukemia-2H3 (RBL-2H3) /CMC-HPLC-ESI-IT-ToF-MS, a procedure has been established to monitor the presence of potential anaphylactic components present in the injection, and the retained compounds were analyzed for their sensitization effect *in vivo* and *in vitro* [77]. A few more such reports were presented in the applications of CMC with different cell membranes such as HMC-1, LAD2, H1R, etc., for the screening of allergenic components present in matrices of therapeutics from plant origin [78–80].

Similarly, osteoplastic active compounds of *Coptidis Rhizoma* were distinguished by incorporating human periodontal ligament cells (hPDLC) in CMC [81]. Neverthless, acetone extract of *Leonurus artemisia* (Lour.) S. Y. Hu (Lamiaceae), which was shown to increase the uterine contraction amplitude, was investigated with Sprague-Dawley rat uterus CMC. The active compound was identified as genkwanin, *in vitro* experiments thereof proved its pharmacological activities [82].

A little different CMC platform was the three-dimensional cell bioreactors. Mou et al. established a 3D cancer cell bioreactor by culturing the cancer cells on a porous scaffold. Live and fixed cells were allowed to interact with the investigated anticancer drugs, as well as non-anticancer drugs. The most productive piece of information from the study was the comparable binding degree differences of these drugs. As the anticancer drugs exhibited a far better degree of binding of >64% than <3% of the non-anticancer drugs, the whole system was potentially applied for the screening of the active compounds from *Polygonum cillinerve* (Nakai) Ohwi (PCO) extract. Online coupling of HPLC-MS enabled the identification of active compounds as aristolochic acid B and aristolochic acid A [83].

Despite its positives, applications of CMC have been restricted due to the negative interactions of biological activity of the cell membrane by the properties of the mobile phase and other stationary phase conditions. Overcoming these drawbacks, an improved version of CMC was presented by Xiao et al. where, the stationary phase equipped with *E. coli* cell membrane which was connected with LC-ToF and applied to screen the antimicrobial peptides from *Jatropha curcas* meal protein isolate hydrolysates (JCMPIH). Comparing the HPLC profiles of filtrate after CMC interaction, with that of JCMPIH, the presence of possible antimicrobial components in the matrix were verified, which were then preceded for their activity and minimum inhibitory concentration [84].

Another important alteration was made to CMC on the fact that incompatible interactions between the bio-membrane and active components, and the modifications of target components can take place in the course of chromatography, which may impair the whole process. Therefore, instead of directly going for chromatography, the natural extract and the bio-membrane were first incubated together in buffer, followed by denaturalization and compound analysis by chromatography-HRMS [85].

Yet another drawback of CMC which is not been reported to resolve somehow, is the restrictions on mobile phase to be adequate for the interactions. While separating the cell membranes, some receptors also get immobilized together, and this can cause the binding of non-targeted ligands also.

## 6.9 SIZE EXCLUSION CHROMATOGRAPHY (SEC)

Size exclusion chromatography (SEC) can easily be defined by its synonym-molecular sieve chromatography. Here, the molecules in a solution are 'sieved' on the basis of their molecular size. Stationary phase consists of porous beads. These pores are of different sizes, and if the compounds coming with the solution have smaller size enough to enter into the pores, they will be trapped inside. The larger molecules will be eluted off easily, leaving the smaller molecules behind to elute later. Generally, it is used for macromolecular systems such as polymers, proteins, etc. Its field of application includes fractionation and characterization of macromolecular complexes, screening of molecular mass distribution, etc., in bio-analytical technologies. SEC is used to separate the ligand-bounded compounds from unbound small molecules (Figure 6.7).

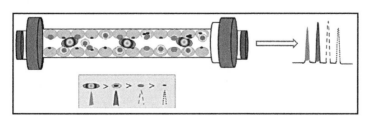

**FIGURE 6.7** Size exclusion chromatography.

SEC was successfully applied for the separation of carbohydrates, heterosides, and aromatic species from the aqueous complex samples produced during the conversion of lignocellulosic biomass into fuels and chemicals. The separated fractions were simultaneously subjected to RPLC-IT-ToF-MS to produce analytical fingerprints and FT-ICR-MS to obtain the exact mass distribution and molecular formulae of compounds [86]. Rather, the application of SEC as a bioanalytical screening technique allowed a simultaneous screening of large number of protein complexes in a single experiment. By using a reproducible SEC-HRMS pipeline, 1693 proteins including 983 cytosolic proteins were identified in two biological replicates from Arabidopsis leaves. Moreover, the study dealt with the quantification of proteins and abundance profile as well as the prediction of their localization and oligomerization states [87]. Another point of view, in which the combination SEC-HRMS is applied, is to determine the molecular mass, molecular mass distribution and degree of polymerization of fructan samples produced out of chicory and ryegrass [88].

SEC separates the compounds solely based on the size, not any other interactions. Hence the molecules should not interact with the stationary phase materials. Rather, the amount of analyte should be kept low, so as the small molecules to occupy completely inside the pores. Moreover, particles of the same size will be eluted together, regardless of their properties.

## 6.10   CONCLUSION

Investigation for healthy and effective medicines is the need of time. While considering the safety part, 'nature made' therapeutics serve well than manmade medicines. Hence, the researches for natural healing compounds are a necessity rather than a world trend. HRMS furnishes its own domain on this consideration. As the concern is to discover bioactive molecules, "bio" analytical tools serve well as the preparatory techniques. Therefore, the chapter discussed the bio-analytical techniques supporting and compatible with HRMS analyzes of phytochemical compounds. There are more techniques and modifications thereof are inventing day-to-day. The only thing that matters is to choose wisely the appropriate technology-instrument combinations, according to the intention of work and the nature of materials involved in the study. Base factor remains the same-HRMS will provide the data.

## KEYWORDS

- **affinity chromatography**
- **cell membrane chromatography**
- **displacement chromatography**
- **HPLC based post-column bioassays**
- **ligand fishing**
- **size exclusion chromatography**
- **TLC bioautography**
- **ultrafiltration**

## REFERENCES

1. Mroczek, T., (2009). Highly efficient, selective, and sensitive molecular screening of acetylcholinesterase inhibitors of natural origin by solid-phase extraction-liquid chromatography/electrospray ionization-octopole-orthogonal acceleration time-of-flight-mass spectrometry and novel thin-layer chromatography-based bioautography. *J. Chromatogr. A, 1216*(12), 2519–2528.
2. Xiao, G., Li, X., Wu, T., Cheng, Z., Tang, Q., & Zhang, T., (2012). isolation of a new meroterpene and inhibitors of nitric oxide production from *Psoralea corylifolia* fruits guided by TLC bioautography. *Fitoterapia, 83*(8), 1553–1557.
3. Bag, A., Bhattacharyya, S. K., Pal, N. K., & Chattopadhyay, R. R., (2012). *In vitro* antibacterial potential of *Eugeniajambolana* seed extracts against multidrug-resistant human bacterial pathogens. *Microbiol. Res., 167*(6), 352–357.
4. Zhang, L., Tang, J., Cheng, Z., Lu, X., Kong, Y., & Wu, T., (2017). Direct coupling of thin-layer chromatography-bioautography with electrostatic field induced spray ionization-mass spectrometry for separation and identification of lipase inhibitors in lotus leaves. *Anal. Chim. Acta., 967*, 52–58.
5. Queiroz, M. M. F., Queiroz, E. F., Zeraik, M. L., Marti, G., Favre-Godal, Q., Simões-Pires, C., Marcourt, L., et al., (2014). Antifungals and acetylcholinesterase inhibitors from the stem bark of *Croton heliotropiifolius,Phytochem. Lett., 10*,lxxxviii–xciii.
6. Mo'ricz, A. M., Ott, P. G., & Morlock, G. E., (2018). Discovered acetylcholinesterase inhibition and antibacterial activity of polyacetylenes in tansy root extract via effect-directed chromatographic fingerprints. *J. Chromatogr. A, 1543*, 73–80.

7. Jamshidi-Aidji, M., & Morlock, G. E., (2015). Bioprofiling of unknown antibiotics in herbal extracts: Development of a streamlined direct bioautography using *Bacillus subtilis* linked to mass spectrometry. *J. Chromatogr. A, 1420,* 110–118.

8. Bhadane, B. S., & Patil, R. H., (2017). isolation, purification, and characterization of antioxidative steroid derivative from methanolic extract of *Carissa carandas* (L.) leaves. *Biocatal. Agric. Biotechnol., 10,* 216–223.

9. Mogashoa, M. M., (2017). *Isolation and Characterization of Antifungal and Antibacterial Compounds from Combretum Molle (Combretaceae) Leaf Extracts.* Doctoral dissertation, University of Pretoria, South Africa.

10. Azadniya, E., & Morlock, G. E., (2017). Bioprofiling of *Salvia miltiorrhiza* via planar chromatography linked to (bio)assays, high resolution mass spectrometry and nuclear magnetic resonance spectroscopy. *J. Chromatogr. A, 1533,* 180–192.

11. Dash, U. C., & Sahoo, A. K., (2017). In vitro antioxidant assessment and a rapid HPTLC bioautographic method for the detection of anticholinesterase inhibitory activity of *Geophila* repens. *J. Integr. Med., 15*(3), 231–241.

12. Dash, U. C., Kanhar, S., Dixit, A., Dandapat, J., & Sahoo, A. K., (2019). Isolation, identification, and quantification of pentylcurcumene from *Geophilarepens*: A new class of cholinesterase inhibitor for Alzheimer's disease. *Bioorg. Chem., 88,* 102947.

13. Benzekri, R., Bouslama, L., Papetti, A., Snoussi, M., Benslimene, I., Hamami, M., & Limam, F., (2016). Isolation and identification of an antibacterial compound from *Diplotaxisharra* (Forssk.) boiss. *Ind. Crops Prod., 80,* 228–234.

14. Jamshidi-Aidji, M., & Morlock, G. E., (2015). Bioprofiling of unknown antibiotics in herbal extracts: Development of a streamlined direct bioautography using *Bacillus subtilis* linked to mass spectrometry. *J. Chromatogr. A, 1420,* 110–118.

15. Van, B. R. B., Huang, C. R., Nikolic, D., Woodbury, C. P., Zhao, Y. Z., & Venton, D. L., (1997). Pulsed ultrafiltration mass spectrometry: A new method for screening combinatorial libraries. *Anal. Chem., 69*(11), 2159–2164.

16. Choi, Y., Jermihov, K., Nam, S. J., Sturdy, M., Maloney, K., Qiu, X., Chadwick, L. R., et al., (2011). Screening natural products for inhibitors of quinone reductase-2 using ultrafiltration LC-MS. *Anal. Chem., 83*(3), 1048–1052.

17. Li, H., Song, F., Xing, J., Tsao, R., Liu, Z., & Liu, S., (2009). Screening and structural characterization of α-glucosidase inhibitors from hawthorn leaf flavonoids extract by ultrafiltration. LC-DAD-MS[n] and SORI-CID FTICR MS. *J. Am. Soc. Mass Spectrom., 20*(8), 1496–1503.

18. Yang, Z., Zhang, Y., Sun, L., Wang, Y., Gao, X., & Cheng, Y., (2012). An ultrafiltration high-performance liquid chromatography coupled with diode array detector and mass spectrometry approach for screening and characterizing tyrosinase inhibitors from mulberry leaves. *Analytica Chimica Acta, 719,* 87–95.

19. Wang, Z., Hwang, S. H., & Lim, S. S., (2017). Comprehensive profiling of minor tyrosinase inhibitors from *Gastrodia elata* using an offline hyphenation of ultrafiltration, high-speed countercurrent chromatography, and high-performance liquid chromatography. *J. Chromatogr. A, 1529,* 63–71.

20. De-qiang, L., Zhao, J., Wu, D., & Shao-Ping, L., (2016). Discovery of active components in herbs using chromatographic separation coupled with online bioassay. *J. Chromatogr. A, 1021,* 81–90.

21. Song, H. P., Zhang, H., Fu, Y., Mo, H. Y., Zhang, M., Chen, J., & Li, P., (2014). Screening for selective inhibitors of xanthine oxidase from flos chrysanthemum using ultrafiltration LC-MS combined with enzyme channel blocking. *J. Chromatogr. B., 961*, 56–61.

22. Jonker, N., Kool, J., Irth, H., & Niessen, W. M., (2011). Recent developments in protein-ligand affinity mass spectrometry. *Anal. Bioanal. Chem., 399*(8), 2669–2681.

23. Liu, J., Burdette, J. E., Xu, H., Gu, C., Van, B. R. B., Bhat, K. P., Booth, N., et al., (2001). Evaluation of estrogenic activity of plant extracts for the potential treatment of menopausal symptoms. *J. Agric. Food Chem., 49*(5), 2472–2479.

24. Nikolic, D., Habibi-Goudarzi, S., Corley, D. G., Gafner, S., Pezzuto, J. M., & Van, B. R. B., (2000). Evaluation of cyclooxygenase-2 inhibitors using pulsed ultrafiltration mass spectrometry. *Anal. Chem., 72*(16), 3853–3859.

25. Cao, H., Yu, R., Choi, Y., Ma, Z. Z., Zhang, H., Xiang, W., Lee, D. Y. W., et al., (2010). Discovery of cyclooxygenase inhibitors from medicinal plants used to treat inflammation. *Pharmacol. Res., 61*(6), 519–524.

26. Cieśla, Ł., & Moaddel, R., (2016). Comparison of analytical techniques for the identification of bioactive compounds from natural products. *Nat. Prod. Rep., 33*(10), 1131–1145.

27. Zhu, Y. T., Jia, Y. W., Liu, Y. M., Liang, J., Ding, L. S., & Liao, X., (2014). Lipase ligands in *Nelumbo nucifera* leaves and study of their binding mechanism. *J. Agric. Food Chem., 62*(44), 10679−10686.

28. Vanzolini, K. L., Jiang, Z., Zhang, X., Vieira, L. C. C., Corrêa, A. G., Cardoso, C. L., Cass, Q. B., & Moaddel, R., (2013). Acetylcholinesterase immobilized capillary reactors coupled to protein coated magnetic beads: A new tool for plant extract ligand screening. *J. Med. Chem., 56*(5), 2038–2044.

29. Wubshet, S. G., Brighente, I. M. C., Moaddel, R., & Staerk, D., (2015). Magnetic ligand fishing as a targeting tool for HPLC-HRMS-SPENMR: α-glucosidase inhibitory ligands and alkylresorcinol glycosides from *Eugenia catharinae*. *J. Nat. Prod., 78*(11), 2657–2665.

30. Deng, S., Xia, L., & Xiao, H., (2014). Screening of a-glucosidase inhibitors from green tea extracts using immobilized enzymes affinity capture combined with UHPLC-Qfig MS analysis. *Chem. Commun., 50*, 2582–2584.

31. Zhu, J., Yi, X., Liu, W., Xu, Y., Chen, S., & Wu, Y., (2017). Immobilized fusion protein affinity chromatography combined with HPLC–ESI-Q-TOF-MS/MS for rapid screening of PPARγ ligands from natural products. *Talanta, 165*, 508–515.

32. Tao, Y., Chen, Z., Zhang, Y., Wang, Y., & Cheng, Y., (2013). Immobilized magnetic beads based multi-target affinity selection coupled with high performance liquid chromatography-mass spectrometry for screening anti-diabetic compounds from a Chinese medicine "Tang-Zhi-Qing." *J. Pharm. Biomed. Anal., 78*, 190–201.

33. Cao, J., Xu, J. J., Liu, X. G., Wang, S. L., & Peng, L. Q., (2016). Screening of thrombin inhibitors from phenolic acids using enzyme-immobilized magnetic beads through direct covalent binding by ultrahigh-performance liquid chromatography coupled with quadrupole time-of-flight tandem mass spectrometry. *J. Chromatogr. A, 1468*, 86–94.

34. Ingkaninan, K., De Best, C. M., Van, D. H. R., Hofte, A. J. P., Karabatak, B., Irth, H., Tjaden, U. R., et al., (2000). High-performance liquid chromatography with

online coupled UV, mass spectrometric and biochemical detection for identification of acetylcholinesterase inhibitors from natural products. *J. Chromatogr. A, 872*(1/2) 61–73.

35. Kchaou, M., Salah, H. B., Mhiri, R., & Allouche, N., (2016). Anti-oxidant and anti-acetylcholinesterase activities of *Zygophyllum album*. *Bangladesh J. Pharmacol., 11,* 54–62.

36. Schobel, U., Frenay, M., Van, E. D. A., McAndrews, J. M., Long, K. R., Olson, L. M., Bobzin, S. C., & Irth, H., (2001). High resolution screening of plant natural product extracts for estrogen receptor alpha and beta binding activity using an online HPLC-MS biochemical detection system. *J. Biomol. Screen, 6*(5), 291–303.

37. De Jong, C. F., Derks, R. J., Bruyneel, B., Niessen, W., & Irth, H., (2006). High-performance liquid chromatography-mass spectrometry-based acetylcholinesterase assay for the screening of inhibitors in natural extracts. *J. Chromatogr. A, 1112*(1/2), 303–310.

38. Potterat, O., & Hamburger, M., (2013). Concepts and technologies for tracking bioactive compounds in natural product extracts: Generation of libraries, and hyphenation of analytical processes with bioassays. *Nat. Prod. Rep., 30,* 546–564.

39. Wang, W., Jiao, L., Tao, Y., Shao, Y., Wang, Q., Yu, R., Mei, L., & Dang, J., (2019). Online HPLC-DPPH bioactivity-guided assay for isolated of antioxidative phenylpropanoids from Qinghai-Tibet plateau medicinal plant *Lanceatibetica. J. Chromatogr. A, 1106,* 1–10.

40. Li, D. Q., Zhao, J., & Li, S. P., (2014). High-performance liquid chromatography coupled with post-column dual-bioactivity assay for simultaneous screening of xanthine oxidase inhibitors and free radical scavengers from complex mixture. *J. Chromatogr. A, 1345,* 50–56.

41. Mocan, A., Schafberg, M., Crişan, G., & Rohn, S., (2016). Determination of lignans and phenolic components of *Schisandra chinensis* (Turcz.) Baill. using HPLC-ESI-ToF-MS and HPLC-online TEAC: Contribution of individual components to overall antioxidant activity and comparison with traditional antioxidant assays. *J. Funct. Foods, 24,* 579–594.

42. Li, S. Y., Yu, Y., & Li, S. P., (2007). Identification of antioxidants in essential oil of radix *Angelicae sinensis* using HPLC coupled with DAD-MS and ABTS-based assay. *Journal of Agricultural andFood Chem., 55*(9), 3358–3362.

43. Inoue, K., Baba, E., Hino, T., & Oka, H., (2012). A strategy for high-speed countercurrent chromatography purification of specific antioxidants from natural products based on online HPLC method with radical scavenging assay. *Food Chem., 134*(4), 2276–2282.

44. Zhu, J., Yi, X., Zhang, J., Chen, S., & Wu, Y., (2017). Chemical profiling and antioxidant evaluation of Yangxinshi tablet by HPLC–ESI-Q-TOF-MS/MS combined with DPPH assay. *J. Chromatogr. A, 1060,* 262–271.

45. Julianti, T., De Mieri, M., Zimmermann, S., Ebrahimi, S. N., Kaiser, M., Neuburger, M., Raith, M., et al., (2014). HPLC-based activity profiling for antiplasmodial compounds in the traditional Indonesian medicinal plant *Carica papaya* L. *J. Ethnopharmacol., 155*(1), 426–434.

46. Van, E. D. A., Schobel, U. P., Lansky, E. P., Irth, H., Van, D., & Greef, J., (2004). Rapid dereplication of estrogenic compounds in pomegranate (*Punica granatum*)

using online biochemical detection coupled to mass spectrometry. *Phytochemistry, 65*(2), 233–241.

47. Song, Z., Wang, H., Ren, B., Zhang, B., Hashi, Y., & Chen, S., (2013). Online study of flavonoids of *Trollius chinensis* Bunge binding to DNA with ethidium bromide using a novel combination of chromatographic, mass spectrometric and fluorescence techniques. *J. Chromatogr. A, 1282*, 102–112.

48. Li, D. Q., Zhao, J., & Li, S. P., (2014). High-performance liquid chromatography coupled with post-column dual-bioactivity assay for simultaneous screening of xanthine oxidase inhibitors and free radical scavengers from complex mixture. *J. Chromatogr. A, 1345*, 50–56.

49. Malherbe, C. J., Willenburg, E., De Beer, D., Bonnet, S. L., Van, D. W. J. H., & Joubert, E., (2014). Iriflophenone-3-C-glucoside from *Cyclopia genistoides*: Isolation and quantitative comparison of antioxidant capacity with mangiferin and isomangiferin using online HPLC antioxidant assays. *J. Chromatogr. A, 951*, 164–171.

50. Giera, M., Heus, F., Janssen, L., Kool, J., Lingeman, H., & Irth, H., (2009). Micro fractionation revisited: A 1536 well high-resolution screening assay. *Anal. Chem., 81*, 5460–5466.

51. Danz, H., Stoyanova, S., Wippich, P., Brattström, A., & Hamburger, M., (2001). Identification and isolation of the cyclooxygenase-2 inhibitory principle in *Isatis tinctoria. Planta Med., 67*(5), 411–416.

52. Kim, H. J., Baburin, I., Khom, S., Hering, S., & Hamburger, M., (2008). HPLC-based activity profiling approach for the discovery of gabaa receptor ligands using an automated two-microelectrode voltage clamp assay on *Xenopus oocytes.Planta Med., 74*, 521–526.

53. Gathungu, R. M., Oldham, J. T., Bird, S. S., Lee-Parsons, C. W. T., Vouros, P., & Kautz, R., (2012). Application of an integrated LC-UV-MS-NMR platform to the identification of secondary metabolites from cell cultures: Benzophenanthridine alkaloids from elicited *Eschscholzia californica* (california poppy) cell cultures. *Anal. Methods, 4*, 1315–1325.

54. Crawford, A., Bohni, N., Maes, J., Kamuhabwa, A., Moshi, M., Esguerra, C., De Witte, P., & Wolfender, J., (2010). Zebrafish bioassay-guided micro fractionation combined with CapNMR a comprehensive approach for the identification of anti-inflammatory and anti-angiogenic constituents of *Rhynchosiaviscosa. Planta Med., 76*(12), 544.

55. Jonker, W., Zwart, N., Stöckl, J. B., De Koning, S., Schaap, J., Lamoree, M. H., Somsen, G. W., et al., (2017). Continuous fraction collection of gas chromatographic separations with parallel mass spectrometric detection applied to cell-based bioactivity analysis. *Talanta, 168*, 162–167.

56. De-qiang, L., Zhao, J., Wu, D., & Shao-ping, L., (2016). Discovery of active components in herbs using chromatographic separation coupled with online bioassay. *J. Chromatogr. A, 1021*, 81–90.

57. Gao, J. M., Kamnaing, P., Kiyota, T., Watchueng, J., Kubo, T., Jarussophon, S., & Konishi, Y., (2008). One-step purification of palmatine and its derivative dl-tetrahydropalmatine from *Enantia chlorantha* using high-performance displacement chromatography. *J. Chromatogr. A, 1208*(1/2), 47–53.

58. Wang, L., Zhao, Y., Zhang, Y., Zhang, T., Kool, J., Somsen, G. W., Wang, Q., & Jiang, Z., (2018). Online screening of acetylcholinesterase inhibitors in natural products using monolith-based immobilized capillary enzyme reactors combined with liquid chromatography-mass spectrometry. *J. Chromatogr. A, 1563*, 135–143.

59. Luo, H., Chen, L., Li, Z., Ding, Z., & Xu, X., (2003). Frontal immunoaffinity chromatography with mass spectrometric detection: A method for finding active compounds from traditional Chinese herbs. *Anal. Chem., 75*(16), 3994–3998.

60. Zhu, J., Yi, X., Liu, W., Xu, Y., Chen, S., & Wu, Y., (2017). Immobilized fusion protein affinity chromatography combined with HPLC–ESI-Q-TOF-MS/MS for rapid screening of PPARγ ligands from natural products. *Talanta, 165*, 508–515.

61. Gao, X., Yang, L., Bai, Y., Li, Q., Zhao, X., Bian, L., & Zheng, X., (2019). Screening of bioactive components from traditional Chinese medicine by immobilized β2 adrenergic receptor coupled with high performance liquid chromatography/mass spectrometry. *J. Chromatogr. B., 1134*, 121782.

62. Zhang, Z., Zhang, Y., Zhao, H., Huang, F., Zhang, Z., & Lin, W., (2017). The important functionality of 14-3-3 isoforms in rice roots revealed by affinity chromatography. *J. Proteom., 158*, 20–30.

63. Oztekin, A., Almaz, Z., Gerni, S., Erel, D., Kocak, S. M., Sengül, M. E., & Ozdemir, H., (2019). Purification of peroxidase enzyme from radish species in fast and high yield with affinity chromatography technique. *J. Chromatogr. A, 1114*, 86–92.

64. Hou, X., Wang, S., Zhang, T., Ma, J., Zhang, J., Zhang, Y., Lu, W., et al., (2014). Recent advances in cell membrane chromatography for traditional Chinese medicines analysis. *J. Pharm. Biomed. Anal., 101*, 141–150.

65. Chen, Y., Zhang, N., Ma, J., Zhu, Y., Wang, M., Wang, X., & Zhang, P., (2016). A Platelet/CMC coupled with offline UPLC-QTOF-MS/MS for screening antiplatelet activity components from aqueous extract of danshen. *J. Pharm. Biomed. Anal., 117*, 178–183.

66. Wang, X., Xue, H., Yue, Y., Xu, D. D., & Peng, X. L., (2015). Targeted screening of active ingredients from *Corydlis decumbentis* rhizoma based on high expression β1 adrenergic receptor/cell membrane chromatography method. *Chin. J. of Experimental Tradi. Medi. Formu., 18*, 65–68.

67. Cao, Y., Chen, X. F., Lü, D. Y., Dong, X., Zhang, G. Q., & Chai, Y. F., (2011). Using cell membrane chromatography and HPLC-TOF/MS method for *in vivo* study of active components from roots of *Aconitum carmichaeli. J. Pharm. Anal., 1*(2), 125–134.

68. Chen, X., Cao, Y., Lv, D., Zhu, Z., Zhang, J., & Chai, Y., (2012). Comprehensive two-dimensional HepG2/cell membrane chromatography/monolithic column/time-of-flight mass spectrometry system for screening antitumor components from herbal medicines. *J. Chromatogr. A, 1242*, 67–74.

69. Chen, X., Cao, Y., Zhang, H., Zhu, Z., Liu, M., Liu, H., Ding, X., et al., (2014). Comparative normal/failing rat myocardium cell membrane chromatographic analysis system for screening specific components that counteract doxorubicin-induced heart failure from *Acontium carmichaeli. Anal. Chem., 86*(10), 4748–4757.

70. Gu, Y., Chen, X., Wang, R., Wang, S., Wang, X., Zheng, L., Zhang, B., et al., (2019). Comparative two-dimensional HepG2 and L02/cell membrane chromatography/C18/

time-of-flight mass spectrometry for screening selective anti-hepatoma components from *Scutellariae radix. J. Pharm. Biomed. Anal., 164*, 550–556.

71. Ji, S. G., Ding, X., & Cao, Y., (2014). Screening of potential active anticancer components of *Bruceajavanica* by SMMC-7721 and Hep-G2 comprehensive two-dimensional CMC-monolith chromatography. *J. Pharm. Pract., 32*, 425–427.

72. Wang, S., Sun, M., Zhang, Y., Du, H., & He, L., (2010). A new A431/cell membrane chromatography and online high-performance liquid chromatography/mass spectrometry method for screening epidermal growth factor receptor antagonists from radix *Sophorae flavescentis. J. Chromatogr. A, 1217*(32), 5246–5252.

73. Li, M., Wang, S., Zhang, Y., & He, L., (2010). An online-coupled cell membrane chromatography with LC/MS method for screening compounds from *Aconitum carmichaeli* debx. acting on VEGFR-2. *J. Pharmaceut. Biomed., 53*(4), 1063–1069.

74. Lv, Y. N., Fu, J., Kong, L. Y., Han, S. L., Zhang, T., & He, L. C., (2017). EGFR antagonism of the irisflorentin from *Rhizoma belamcandae* using CMC-Online-HPLC-IT-TOF MS system. *J. Chinese Mass Spectrom. Soc., 4*, 8.

75. Han, S., Li, C., Huang, J., Wei, F., Zhang, Y., & Wang, S., (2016). Cell membrane chromatography coupled with UHPLC–ESI–MS/MS method to screen target components from *Peucedanumpraeruptorum* Dunn acting on α1A adrenergic receptor. *J. Chromatogr. A, 1011*, 158–162.

76. Zhang, T., Han, S., Huang, J., & Wang, S., (2013). Combined fibroblast growth factor receptor 4 cell membrane chromatography online with high performance liquid chromatography/mass spectrometry to screen active compounds in *Brassica albla. J. Chromatogr. A, 912*, 85–92.

77. Han, S., Lv, Y., Xue, W., Cao, J., Cui, R., & Zhang, T., (2016). Screening anaphylactic components of MaiLuoNing injection by using rat basophilic leukemia-2H3 cell membrane chromatography coupled with HPLC-ESI-TOF-MS. *J. Sep. Sci., 39*(3), 466–472.

78. Lin, Y., Wang, C., Hou, Y., He, H., Huang, L., Yang, L., & Sun, M., (2017). The human mast cell line-1 cell membrane chromatography coupled with HPLC-ESI-MS/MS method for screening potentical anaphylactic components from Chuan Xin Lian injection. *Biomed. Chromatogr., 31*(12), e4015.

79. Guo, Y., Han, S., Cao, J., Liu, Q., & Zhang, T., (2014). Screening of allergic components mediated by H1R in homoharringtonine injection through H1R/CMC-HPLC/MS. *Biomed. Chromatogr., 28*(12), 1607–1614.

80. Lv, Y., Fu, J., Shi, X., Yang, Z., & Han, S., (2017). Screening allergic components of Yejuhua injection using LAD2 cell membrane chromatography model online with high performance liquid chromatography-ion trap-time of flight-mass spectrum system. *J. Chromatogr. A, 1055*, 119–124.

81. Liu, J., Yang, J., Wang, S., Sun, J., Shi, J., Rao, G., Li, A., & Gou, J., (2012). Combining human periodontal ligament cell membrane chromatography with online HPLC/MS for screening osteoplastic active compounds from *Coptidis rhizoma. J. Chromatogr. A., 904*, 115–120.

82. Fan, J., Wei, F., Zhang, Y., Su, H., Ji, Z., He, J., & Han, S., (2016). Combining Sprague-Dawley rat uterus cell membrane chromatography with HPLC/MS to screen active components from *Leonurus artemisia. Pharm. Biol., 54*(2), 279–284.

83. Mou, Z. L., Qi, X. N., Liu, R. L., Zhang, J., & Zhang, Z. Q., (2012). Three-dimensional cell bioreactor coupled with high performance liquid chromatography-mass spectrometry for the affinity screening of active components from herb medicine. *J. Chromatogr. A., 1243*, 33–38.

84. Xiao, J., & Zhang, H., (2012). An Escherichia coli cell membrane chromatography-offline LC-TOF-MS method for screening and identifying antimicrobial peptides from *Jatropha curcas* meal protein isolate hydrolysates. *J. Biomol. Screen., 17*(6), 752–760.

85. Li, S., Zhao, J., Qian, Z., & LI, J., (2010). Advanced development of chromatography in screening and identification of effective compounds in Chinese *Materia medica. Sci. Sin. Chim., 40*(6), 651–667.

86. Dubuis, A., Le Masle, A., Chahen, L., Destandau, E., & Charon, N., (2020). Offline comprehensive size exclusion chromatography× reversed-phase liquid chromatography coupled to high-resolution mass spectrometry for the analysis of lignocellulosic biomass products. *J. Chromatogr. A, 1609*, 460505.

87. Aryal, U. K., McBride, Z., Chen, D., Xie, J., & Szymanski, D. B., (2017). Analysis of protein complexes in *Arabidopsis* leaves using size exclusion chromatography and label-free protein correlation profiling. *[J. Proteom., 166*, 8–18.

88. Evans, M., Gallagher, J. A., Ratcliffe, I., & Williams, P. A., (2016). Determination of the degree of polymerization of fructans from ryegrass and chicory using MALDI-TOF mass spectrometry and gel permeation chromatography coupled to multiangle laser light scattering. *Food Hydrocoll., 53*, 155–162.

# CHAPTER 7

# Plant Metabolites, While Looking Through HRMS: Characterization of the Phenolic Profile of *Lactuca sativa* as a Case Study

GABRIELA E. VIACAVA,[1] LUIS A. BERRUETA,[2] BLANCA GALLO,[2] and ROSA M. ALONSO-SALCES[3]

[1]*Research Group of Food Engineering, National Council of Scientific and Technological Research (CONICET), Department of Chemistry and Food Engineering, Faculty of Engineering, National University of Mar del Plata, 4302 Juan B. Justo Street, Mar del Plata – 7600, Argentina*

[2]*Department of Analytical Chemistry, Faculty of Science and Technology, University of the Basque Country/Euskal Herriko Unibertsitatea (UPV/EHU), PO Box – 644, 48080, Bilbao, Spain*

[3]*Research Group of Applied Microbiology, Social Bee Research Center, Institute for Research in Production, Health and Environment, CONICET, Department of Biology, Faculty of Exact and Natural Sciences, National University of Mar del Plata, Funes – 3350, Mar del Plata – 7600, Argentina*

## ABSTRACT

Higher plants synthesize a wide range of secondary metabolites, such as phenolics, terpenoids, and alkaloids, which contribute to odor, taste, and color to the plants and are involved in protection mechanisms against biotic and abiotic stresses. These compounds also have an important role in the food, pharmaceutical, and cosmetic industries, since they are use as antimicrobials, antioxidants, coloring-agents, flavoring-agents, agrochemicals, and biopesticides, among others. In particular, most of these plant-derived compounds relevant for industrial applications are phenolics (e.g., phenolic acids (PA), flavonoids, coumarins, and lignans). For the investigation of structure-activity relationships (SARs) and quality control (QC), the access to rapid and reliable analytical methods for the identification and determination of these natural compounds is important. Mass spectrometry coupled to liquid chromatography and UV-visible detection is a powerful tool for the analysis of phenolic compounds. Technological advances introduced in the last years have provided improvements in terms of chromatographic separation but particularly in the field of mass spectrometry. Indeed, the emergence of ultrahigh performance liquid chromatography (UHPLC) coupled to high resolution mass spectrometry (HRMS), such as the time-of-flight (ToF) mass spectrometry (ToF/MS) or a quadrupole-ToF (QToF/MS) instruments, made possible to develop a very attractive analytical methodology that allows to perform high resolution and accurate mass measurements of the precursor and fragment ions, providing valuable structural information for irrefutable compound identification.

This chapter aims to present a powerful analytical methodology for the characterization of phenolic compounds in complex plant material. The phenolic profiles of different lettuce cultivars (butterhead, green oak-leaf, and red oak-leaf) were thoroughly studied since this leafy vegetable is one of the most popular in the world, constituting a major dietary source of phenolic compounds. The analytical strategy included the use of UHPLC coupled online to diode array detection (DAD), electrospray ionization interface (ESI), and QToF/MS. MS$^E$ acquisition mode was also used to maximize the QToF instrument duty cycle by collecting automatic and simultaneous information of exact mass at high and low collision energies of precursor ions, as well as other ions produced as a result of their fragmentation, over a single experimental run. One hundred seventeen phenolic compounds were identified in the acidified hydromethanolic extract of freeze-dried lettuce leaves: 40 hydroxycinnamic acid derivatives, 21 hydroxybenzoic acid derivatives, 2 hydroxyphenylacetic acid derivatives, 18 flavonols, 11 flavones, 1 flavanone, 4 anthocyanidins, 7 coumarins, 1 hydrolysable tannin, and 12 lignans. Forty-eight of these compounds were tentatively identified for the first time in lettuce in the present study, and only 20 of them had been previously reported in oak-leaf lettuce cultivars in literature. Moreover, the phenolic profile of the butterhead lettuce cultivar had not been described before. The UHPLC-DAD-ESI-QToF/MS$^E$ approach provided new structural information and allowed the identification of unknown phenolics demonstrating to be a useful tool for the characterization of phenolic compounds in complex plant matrices.

## 7.1 INTRODUCTION

### 7.1.1 PLANT METABOLITES

Plants, being sessile organisms, produce thousands of metabolites which are small molecules with molecular weights (MW) of <1000 Da with specialized functions. Primary metabolites, including amino acids, lipids, and carbohydrates, constitutively accumulate in plant cells and are essential to sustain life via normal metabolic processes. Higher plants also synthesize a wide range of secondary metabolites, such as phenolics, alkaloids, and terpenes, which are not essential for plant growth, development, or reproduction, but play a recognized role in the interaction of the plant with its environment [1]. Natural functions of these latter compounds, also referred to as natural products, include protection against UV radiation, scavenging of radicals, defense against phytopathogenic bacteria, fungi or viruses, or attraction of pollinators [2]. In addition, secondary metabolites synthesized in plants are the best sources of chemical diversity and have an important role in the food, pharmaceutical, and cosmetic industries, since they are used as antimicrobials, antioxidants, coloring-agents, flavoring-agents, agrochemicals, and biopesticides, among others [3]. Most of the plant-derived compounds relevant for industrial applications are phenolics (e.g., phenolic acids (PA), flavonoids, coumarins, and lignans), whose analysis is the topic of this chapter.

### 7.1.2 PHENOLIC COMPOUNDS

Among secondary metabolites, phenolic compounds are one of the most widespread and diverse group, with more than 8,000 phenolic structures currently known. These compounds are derived from the shikimate pathway and phenylpropanoid metabolism (Figure 7.1) and are considered one of the most important classes of secondary metabolites. They are involved in protection mechanisms of the plant against biotic and abiotic stresses, but also contribute to its sensorial properties, such as color, astringency, bitterness, aroma, and flavor, as well as to the preservation of food industry products [4]. Moreover, the intake of phenolic compounds through fruits and vegetables has been proved to provide beneficial effects against oxidative stress, cancer, and cardiovascular diseases, among others, due to their anti-inflammatory and antioxidant properties [5]. As a consequence, this kind of compounds plays a decisive role in the nutritional, organoleptic, and commercial properties of plant-derived products and has gained a great relevance in human diet.

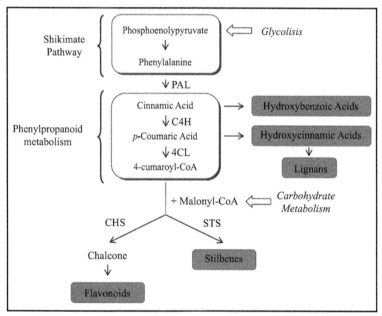

**FIGURE 7.1**　Biosynthetic pathway for plant-derived phenolic compounds. (*Abbreviations:*PAL: phenylalanina ammonia-lyase, C4H: cinnamate 4-hydroxylase, 4CL: 4-coumaroyl-CoA-ligase; CHS: chalcone synthase; STS: stilbene synthase).

*Source:* Adapted from: Winkel-Shirley [57].

The structural unit of phenolic compounds consists of a phenol (an aromatic ring bearing at least one hydroxyl substituent), and can be classified into several classes according to the number of phenol rings that they contain and to the structural elements that bind these rings to one another. The main groups of phenolics are: PA (derivatives of cinnamic acid or benzoic acid), flavonoids, tannins (hydrolysable and condensed), stilbenes, and lignans. In addition to this diversity, phenolics may be associated with mono- or polysaccharides (glycosides), in which one or more hydroxyl groups are bounded to a sugar with formation of an acid labile glycosidic O-C bond, or functional derivatives such as esters or methyl esters. The associated sugar moiety is very often glucose or rhamnose, but other sugars may also be involved (e.g., galactose, arabinose, xylose, glucuronic acid) [6]. Disaccharides are also often found in association with flavonoids, the most common ones being rutinose (rhamnosyl-($\alpha$1→6)-glucose) and neohesperidose (rhamnosyl-($\alpha$1→2)-glucose), and occasionally, tri-, and even tetrasaccharides are encountered [7].

## 7.1.2.1 *PHENOLIC COMPOUNDS IN LETTUCE*

Given the important biological role of phenolic compounds in plant metabolism and their invaluable uses in multiple industries, numerous studies have been performed in order to determine the phenolic profile of different vegetable tissues. Researches carried out to characterized phenolics in leafy vegetables are not an exception. Among these vegetables, lettuce (*Lactuca sativa* L.), belonging to the Compositae (*Asteraceae*) family, is of particular interest since it is one of the most popular food crop in the world, commonly consumed fresh in salad dishes. Furthermore, lettuce is a basic ingredient for the industry of minimally processed food products and ranks highly both in production and economic value among vegetables [8]. In addition, this vegetable is of particular interest in nutrition because of its phytochemical content, including phenolics with antioxidant properties, associated with human health benefits as mentioned above [9–11]. Several studies showed the health effect of lettuce consumption in improving the lipid status and preventing tissue lipid peroxidation in rats, and increasing plasma total antioxidant capacity and antioxidant levels in humans [12, 13]. However, lettuce is not the richest vegetable in phenolics, a considerable amount of daily phenolic intake is provided by it because of its high consumption all over the world [14]. The main classes of phenolic compounds found in different varieties of lettuce are PA and flavonoids [15–17]. However, the contents of phenolics in lettuce tissues are susceptible to high variations among cultivars and growing conditions, and many compounds still remain unidentified [18–20]. Therefore, further research to characterize the phenolic profiles of different lettuce cultivars is fully justified.

## 7.1.3 ANALYSIS OF PHENOLIC COMPOUNDS

In the last decades, many efforts have been made in order to provide highly sensitive and selective analytical methods for the determination and characterization of phenolics in plant materials.

Reverse-phase high-performance liquid chromatography (RP-HPLC) is the main method used for the separation of phenolic compounds in plant-food material. The chromatographic conditions include the use of, almost exclusively, a reversed-phase silica-bonded C18 column, UV-visible absorbance detection, and a gradient elution using a binary solvent system containing high quality ultrapure acidified water (phosphoric, acetic, formic acids) and a less polar organic solvent (acetonitrile or methanol) [21].

The demands of high sample throughput in short time frames have given rise to the development of an ultrahigh performance liquid chromatography (UHPLC) technology, which operate with smaller particle sizes (<2 μm) of the stationary phase and at higher pressures (up to 1300 mbar). UHPLC systems have allowed higher separation efficiency and shorter analysis times as well as increases in peak capacity, sensitivity, and reproducibility compared to conventional HPLC [22].

The diode array detector (DAD) provides complementary information for the identification of the main phenolic structures present in plants since each phenolic class present characteristic UV-visible spectra. However, liquid chromatography coupled to high-resolution mass spectrometry has become the best tool to confirm the identity of target compounds [7].

Mass spectrometry can achieve very high sensitivity and provide information on the molecular weight and structural features of the phenolic compound, i.e., the aglycone moiety, the types of carbohydrates (mono-, di-, tri- or tetrasaccharides and hexoses, desoxyhexoses or pentoses) or other types of substituent present in the molecule, the stereochemical assignment of terminal monosaccharide units, the sequence of the glycan part, the interglycosidic linkages, and the substituent position on the aglycone structure [8].

The hybrid quadrupole-time-of-flight mass analyzer (QToF/MS) provides excellent mass accuracy over a wide dynamic range and measurements of the true isotope pattern that elucidates the molecular formula of unknown metabolites with a high degree of reliability [23]. This instrument also performs tandem MS, which is useful as a structural confirmation tool when standard compounds are not available. Furthermore, QToF provides high selectivity by the extracted ion chromatogram (EIC) mode when there are overlapping peaks, where spectrophotometric detection could be limited. Thus, LC coupled to highly sensitive and high-resolution MS, such as QToF, enables the separation and detection of minor compounds that could co-elute and be underestimated vs. major ions, not being identified or even detected by older methodologies. LC coupled to QToF/MS has been applied in targeted analysis but is suitable also for the extensive profiling of hundreds of natural plant metabolites [24]. Among the ionization sources, electrospray ionization (ESI) is extremely useful for the analysis of large, non-volatile, chargeable molecules such as phenolic compounds.

Technological advances such as the so-called MS$^E$ acquisition method maximizes the QToF instrument duty cycle, performing simultaneous collection of precursor ions as well as other ions produced as a result of their fragmentation in exact mass mode over a single experimental run. To achieve this, the spectrometer cycles rapidly and continuously between two functions. The first scan function acquires a wide mass range at low collision energy and collects information about the unfragmented ions in the ionization source. The second scan function acquires data over the same mass range but using higher collision energy (fixed or ramped). This scan function allows the selective collection of fragmented ion data based on all ions acquired in the first scan, which is equivalent to a simultaneous batch of selective tandem mass spectrometric (MS/MS) product ions scans from all precursor ions coming from ion source. The usefulness of MS$^E$ data acquisition mode for the identification of phenolic compounds in complex plant samples in just one injection was demonstrated by Ramirez-Ambrosi et al. [25].

## 7.2 ANALYSIS OF PHENOLIC COMPOUNDS IN LETTUCE

### 7.2.1 CHEMICALS AND PLANT MATERIAL

Water, methanol, acetonitrile, and formic acid (Fisher Scientific, Fair Lawn, NJ, USA) were of Optima® LC/MS grade; ascorbic acid (Panreac, Barcelona, Spain), analytical grade; and glacial acetic acid (Merck, Darmstadt, Germany), Suprapur® quality. Leucine Enkephalin acetate hydrate and sodium formate solution were provided by Sigma-Aldrich Chemie (Steinheim, Germany). Luteolin-7-*O*-glucoside, kaempferol-3-*O*-glucoside, quercetin-3-*O*-galactoside, quercetin-3-*O*-rhamnoside,cyanidin-3-O-glucoside, and cyanidin-3-O-galactoside were purchased from Extrasynthèse (Genay, France); caffeoyltartaric acid and quercetin-3-*O*-glucoside, from Chromadex (Irvine, CA, USA); 5-*O*-caffeoylquinic acid, *p*-coumaric acid, 1,5-dicaffeoylquinic acid, 1,3-dicaffeoylquinic acid, and quercetin-3-*O*-rutinoside, from Sigma-Aldrich Chemie (Steinheim, Germany); and ferulic acid, caffeic acid, and 3,4-dihydroxybenzoic acid, from Fluka Chemie (Steinheim, Germany). Standard stock solutions of phenolic compounds were prepared in methanol, except for anthocyanins which were prepared in methanol-HCl 30% (99:1, v/v). Dilutions from stock solutions were made in methanol-water-acetic acid (30:65:5, v/v/v).

Three cultivars of lettuce (*Lactuca sativa* L.), two green varieties (butterhead and green oak-leaf), and one red variety (red oak-leaf) were studied. Lettuces were provided by a local producer in Sierra de los Padres (Mar del Plata, Argentina), having been grown under identical agronomic field conditions following traditional standard procedures for lettuce cultivation. The leaves of ten lettuce plants of each cultivar were frozen with liquid nitrogen and freeze-dried, homogenized, and crushed to obtain a homogeneous powder, which was stored at room temperature protected from light and humidity in a desiccator until analysis.

### 7.2.2   EXTRACTION OF PHENOLIC COMPOUNDS

Freeze-dried lettuce (0.1 g) was extracted with 5 mL of methanol-water-acetic acid (30:65:5, v/v/v) containing 2 g/L ascorbic acid, to prevent polyphenol oxidation by the polyphenoloxidase enzyme, in an ultrasonic bath for 10 min. Then, the extract was centrifuged at 6000 rpm during 15 min at 4°C, and the supernatant was filtered through a 0.45 μm PTFE filter (Waters, Milford, CA, USA) prior to injection into the UHPLC system. Analyzes were performed in triplicate.

### 7.2.3   UHPLC-DAD-ESI-QTOF/MS$^E$ ANALYSIS

Lettuce extract was analyzed using an ACQUITY UHPLC™ system from Waters (Milford, MA, USA), equipped with a binary solvent delivery pump, an autosampler, a column compartment, a DAD detector, and controlled by MassLynx v4.1 software. A reverse-phase Acquity UHPLC BEH C18 column (2.1 mm × 100 mm, 1.7 μm) and an Acquity UHPLC BEH C18 VanGuard™ pre-column (1.7 μm) from Waters (Milford, USA) were used. The flow rate was 0.5 mL/min; injection volume, 5 μL; column and autosampler temperatures, 40°C and 4°C, respectively. Mobile phases consisted of 0.1% (v/v) acetic acid in water (A) and 0.1% (v/v) acetic acid in methanol (B). The elution conditions applied were: 0–8.5 min, linear gradient 0–13% B; 8.5–11 min, 13% B isocratic; 11–12.3 min, linear gradient 13–15% B; 12.3–13.8 min, linear gradient 15–19% B; 13.8–17.3 min, linear gradient 19–23% B; 17.3–19 min, 23% B isocratic; 19–24 min, linear gradient 23–30% B; 24–26 min, 30% B isocratic; 26–27 min, linear gradient 30–100% B; 27–28 min, 100% B isocratic; and finally reconditioning of the column with 100% A isocratic. UV-visible spectra were recorded from 210 to 500 nm (20 Hz, 1.2 nm resolution). Hydroxybenzoic acids were monitored at 254 nm; flavanones at 280 nm; hydroxycinnamic acids and coumarins at 320 nm; flavonols and flavones at 370 nm; and anthocyanins at 500 nm.

All MS data acquisitions were performed on a SYNAPT™ G2 HDMS with a quadrupole time-of-flight (QToF) configuration (Waters, Milford, MA, USA) equipped with an ESI source operating in both positive and negative modes. The capillary voltage was set to 0.7 kV (ESI+) or 0.5 kV (ESI–). Nitrogen was used as the desolvation and cone gas at flow rates of 900 L/h and 10 L/h, respectively. The source and desolvation temperatures were 120°C and 400°C, respectively. Leucine-enkephalin solution (2 ng/μL) in 0.1% (v/v) formic acid in acetonitrile-water (50:50, v/v) was used for the lock mass correction ($m/z$ 556.2771 and 278.1141, or $m/z$ 554.2615 and 236.1035, depending on the ionization mode, were monitored at scan time 0.2 s, interval 10 s, scans to average 3, mass window ±0.5 Da, cone voltage 30 V, at a flow rate 10 μL/min). Data acquisition was recorded in the mass range 50–1200 $u$ in resolution mode (FWHM ≈ 20,000) with a scan time of 0.2 s and an interscan delay of 0.024 s, and automatically corrected during acquisition based on the lock mass. Before analysis, the mass spectrometer (MS) was mass calibrated with the sodium formate solution. To perform MS$^E$ mode analysis, the cone voltage was set to 20 V (ESI+) or 30 V (ESI–) and the quadrupole operated in a wide band RF mode only. Two discrete and independent interleaved acquisition functions were automatically created. The first function, typically set at 6 eV in trap cell of the T-Wave, collects low energy or unfragmented data while the second function collects high energy or fragmented data typically using 6 eV in trap cell and a collision ramp 10–40 eV in transfer cell. In both cases, Argon gas was used for collision induced dissociation (CID). Data were recorded in continuous mode. For instrument control, data acquisition and processing MassLynxTM software Version 4.1 (Waters MS Technology, Milford, USA) was used.

## 7.2.4  IDENTIFICATION OF PHENOLIC COMPOUNDS

The identification of the phenolic compounds for which standards were available was carried out by the comparison of their retention times (RTs), their UV-vis spectra, and $MS^E$ spectra recorded in positive and negative mode with those obtained by injecting standards in the same conditions. The identity of the rest of compounds was elucidated using the following analytical data: (i) the UV-vis spectrum when it was available to assign the phenolic class, since each class exhibits a characteristic UV-vis spectrum; (ii) the low collision energy $MS^E$ spectrum in positive and negative ion mode to determine the molecular weight; and since only the protonated/ deprotonated molecules are able to form in the ESI source adducts, clusters, and/or molecular complexes with mobile phase species (e.g., adducts with sodium $[M+Na]^+$ at 22 $u$ above the protonated molecule, $[2M+Na]^+$ of monoacyl hydroxycinnamic acids, the dehydrated proton-ated molecule ($[M+H-H_2O]^+$) of PA and diacyl hydroxycinnamic acids in positive mode; and adducts with $HSO_4^-$ (97 $u$) and $AcO^-$ (43 $u$) and the deprotonated dimer ion $[2M-H]^-$ of monoacyl hydroxycinnamic acid in negative mode), their presence in the low collision energy spectra allows the unequivocal identification of the $[M+H]^+$ or $[M-H]^-$ ions; and (iii) the high collision energy $MS^E$ spectrum provides the phenolics fragmenta-tion patterns, which afford structural information related to the type of carbohydrates, the sequence of the glycan part, interglycosidic linkages and the aglycone moiety, allowing to assign the protonated aglycone $[Y_0]^+$ and/or the deprotonated aglycone $[Y_0]^-$. The identification of the aglycone was carried out based on the observation of $^{i,j}A^+$ and $^{i,j}B^+$ ions [7, 26, 27]. Furthermore, the chromatographic elution order aided in some structural assignments, as well as bibliographic references. IUPAC nomenclature and recommended numbering system were used for chlorogenic acids

(CGAs) and flavonoids; and common names were used for other phenolic acid derivatives, coumarins, hydrolysable tannins and lignan derivatives [28]. Structures of each family of compounds studied are presented in Figure 7.2.

## 7.3 PHENOLIC PROFILES OF *LACTUCA SATIVA* SPECIES

In the present study, UHPLC-DAD-ESI-QToF/MS$^E$ was used to identify the phenolic compounds present in butterhead and green and red oak-leaf lettuce cultivars [29, 30]. Table 7.1 shows the UV-visible and MS spectral data used for their tentative identification: 111 phenolic compounds in the butterhead lettuce cultivar, 109 in the green oak leaf lettuce cultivar, and 113 compounds in the red cultivar. Previous studies had characterized up to 95 phenolics in other green lettuce varieties by UHPLC-ESI-ToF/MS and MS/MS and up to 24 phenolics by LC coupled to ion trap (IT) or triple quadrupole (QqQ) mass spectrometry in red varieties [16–18, 23, 31–35].

### 7.3.1 PHENOLIC ACID DERIVATIVES

The identification of phenolic acid derivatives, i.e., hydroxycinnamic, hydroxybenzoic, and hydroxyphenyl acetic derivatives, was carried out taking into account mainly the negative ion mode mass spectra; the positive ion mode being used for verification. In the high collision energy MS spectra, losses of $H_2O$, $CO_2$ and CO were regularly observed, which have also been described by other authors using IT, QqQ, and QToF [25, 36].

#### 7.3.1.1 HYDROXYCINNAMIC DERIVATIVES

#### 7.3.1.1.1 Caffeoylquinic Acids

Three major chromatographic peaks (1, 3, 6), presenting the same UV spectra as the standard *trans*-5-caffeoylquinic acid (*trans*-5-CQA), were detected in the chromatograms extracted from the Total Ion Current (TIC) MS scan chromatogram in negative and positive modes at *m/z* 353 and 355 respectively, which were due to three caffeoylquinic acid (CQA) isomers. Compound 3 (Rt = 7.32 min, λmax = 300, 324 nm) was identified

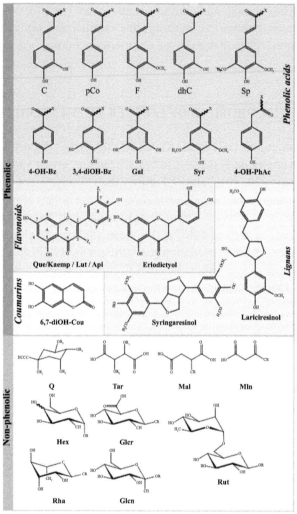

**FIGURE 7.2** Chemical structures of phenolic compounds found in different lettuce cultivars. (*Abbreviations for the phenolic moieties:* C: caffeoyl; pCo: p-coumaroyl; F: feruloyl; dhC: dihydrocaffeoyl; Sp: sinapoyl; 4-OH-Bz: 4-hydroxybenzoyl; 3,4-diOH-Bz: 3,4-dihydroxybenzoyl; Gal: galloyl; Syr: syringoyl; 4-OH-PhAc: 4-hydroxyphenylacetoyl; Que: quercetin ($Z_1$ = OH, $Z_2$ = OH); Kaemp: kaempferol ($Z_1$ = H, $Z_2$ = OH); Lut: luteolin ($Z_1$ = OH, $Z_2$ = H); Api: apigenin ($Z_1$ = H, $Z_2$ = H); 6,7-diOH-Cou: 6,7-dihidroxycoumarin. *Abbreviations for the non-phenolic moieties:* Q: quinic acid; Tar: tartaric acid; Mal: malic acid; Mln: malonic acid; Hex: hexose; Glcr: glucuronic acid; Rut: rutinose (rhamnosylglucose); Rha: rhamnose; Glcn: gluconic acid. R, R1, R2, R3, R4, and R5 in non-phenolic moieties can be esterified in position X of phenolic acids or etherified with phenolic OH groups).

**TABLE 7.1** Retention Times, UV-Visible Maxima and MS$^E$ Data of Polyphenols Identified by UHPLC-DAD-ESI-Q-ToF/MS in Lettuce Varieties [a,b,c]

| Nº | LC Rt (min) | DAD UV bands (nm) | ESI(+)-QToF/MS | | | | ESI(-)-QToF/MS | | | | Assignment |
|---|---|---|---|---|---|---|---|---|---|---|---|
| | | | Exp. Acc. Mass [M+H]+ | Error (mDa) | Formula [M+H]+ | Adducts & fragment ions of [M+H]+ m/z | Exp. Acc. Mass [M-H]- | Error (mDa) | Formula [M-H]- | Adducts & fragment ions of [M-H]- m/z | Tentative identification |
| **Hydroxycinnamic derivatives** | | | | | | | | | | | |
| *Caffeoylquinic acids* | | | | | | | | | | | |
| 1 | 4.74 | 301 sh, 323 | 355.1068 | 3.9 | $C_{16}H_{19}O_9$ | 377.0858 [M+Na]+<br>163.0398 [Caffeoyl+H]+<br>145.0279 [Caffeoyl+H-H₂O]+<br>135.0448 [Caffeoyl+H-CO]+<br>117.0343 [Caffeoyl+H-CO-H₂O]+<br>89.0397 [Caffeoyl+H-H₂O-2CO]+ | 353.0872 | -0.1 | $C_{16}H_{17}O_9$ | 191.0556 [Quin-H]- (100)<br>179.0348 [Caffeic-H]- (32)<br>173.0437 [Quin-H-H₂O]- (4)<br>135.0446 [Caffeic-H-CO₂]- (71) | 3-*trans*-*O*-Caffeoylquinic acid |
| 2 | 6.65 | - | 355.1026 | -0.3 | $C_{16}H_{19}O_9$ | 731.1791 [2M+Na]+<br>551.1234 [2M+Na-caffeic]+<br>377.0846 [M+Na]+<br>163.0421 [Caffeoyl+H]+<br>145.0279 [Caffeoyl+H-H₂O]+<br>135.0433 [Caffeoyl+H-CO]+<br>117.0342 [Caffeoyl+H-CO-H₂O]+<br>89.0396 [Caffeoyl+H-H₂O-2CO]+ | 353.0869 | 0.4 | $C_{16}H_{17}O_9$ | 707.1821 [2M-H]-<br>191.0561 [Quin-H]- (100) | 1-*trans*-*O*-Caffeoylquinic acid |
| 3 | 7.32 | 300 sh, 324 | 355.1026 | -0.3 | $C_{16}H_{19}O_9$ | 731.1791 [2M+Na]+<br>551.1234 [2M+Na-caffeic]+<br>377.0846 [M+Na]+<br>163.0421 [Caffeoyl+H]+<br>145.0279 [Caffeoyl+H-H₂O]+<br>135.0433 [Caffeoyl+H-CO]+<br>117.0342 [Caffeoyl+H-CO-H₂O]+<br>89.0396 [Caffeoyl+H-H₂O-2CO]+ | 353.0869 | -0.4 | $C_{16}H_{17}O_9$ | 707.1821 [2M-H]-<br>191.0556 [Quin-H]- (100)<br>179.0343 [Caffeic-H]- (1)<br>173.0449 [Quin-H-H₂O]- (3)<br>135.0443 [Caffeic-H-CO₂]- (2) | 5-*trans*-*O*-Caffeoylquinic acid |

**TABLE 7.1** (Continued)

| N° | LC Rt (min) | DAD UV bands (nm) | ESI(+)-QToF/MS | | | | ESI(-)-QToF/MS | | | | Assignment |
|---|---|---|---|---|---|---|---|---|---|---|---|
| | | | Exp. Acc. Mass [M+H]+ | Error (mDa) | Formula [M+H]+ | Adducts & fragment ions of [M+H]+ m/z | Exp. Acc. Mass [M-H]- | Error (mDa) | Formula [M-H]- | Adducts & fragment ions of [M-H]- m/z | Tentative identification |
| 4 | 8.12 | – | 355.1068 | 3.9 | $C_{16}H_{19}O_9$ | 731.1739 [2M+Na]+; 709.1981 [2M+H]+; 163.0397 [Caffeoyl+H]+; 145.0128 [Caffeoyl+H-H₂O]+; 135.0463 [Caffeoyl+H-CO]+; 117.0333 [Caffeoyl+H-CO-H₂O]+; 89.0383 [Caffeoyl+H-H₂O-2CO]+ | 353.0861 | -1.2 | $C_{16}H_{17}O_9$ | 707.1796 [2M-H]-; 191.0557 [Quin-H]- (100); 179.0344 [Caffeic-H]- (12); 135.0441 [Caffeic-H-CO₂]- (21) | 3-cis-O-Caffeoylquinic acid |
| 5 | 8.36 | – | 355.1068 | 3.9 | $C_{16}H_{19}O_9$ | 377.0844 [M+Na]+; 163.0445 [Caffeoyl+H]+; 145.0325 [Caffeoyl+H-H₂O]+; 135.0408 [Caffeoyl+H-CO]+; 117.0364 [Caffeoyl+H-CO-H₂O]+ | 353.0865 | -0.8 | $C_{16}H_{17}O_9$ | 191.0554 [Quin-H]- (100); 173.0458 [Quin-H-H₂O]- (13) | 4-trans-O-Caffeoylquinic acid |
| 6 | 10.23 | 301 sh, 316 | 355.1068 | 3.9 | $C_{16}H_{19}O_9$ | 731.1746 [2M+Na]+; 551.1199 [2M+Na−caffeic]+; 377.0841 [M+Na]+; 163.0400 [Caffeoyl+H]+; 145.0284 [Caffeoyl+H-H₂O]+; 135.0443 [Caffeoyl+H-CO]+; 117.0346 [Caffeoyl+H-CO-H₂O]+; 89.0396 [Caffeoyl+H-H₂O-2CO]+ | 353.0867 | -0.6 | $C_{16}H_{17}O_9$ | 707.1816 [2M-H]-; 191.0557 [Quin-H]- (100); 173.0449 [Quin-H-H₂O]- (3) | 5-cis-O-Caffeoylquinic acid |
| 7 | 15.06 | – | | | $C_{16}H_{19}O_9$ | 163.0399 [Caffeoyl+H]+; 145.0287 [Caffeoyl+H-H₂O]+; 135.0446 [Caffeoyl+H-CO]+; 117.0278 [Caffeoyl+H-CO-H₂O]+ | 353.0876 | 0.3 | $C_{16}H_{17}O_9$ | 191.0578 [Quin-H]- (100); 179.0314 [Caffeic-H]- (5); 173.0455 [Quin-H-H₂O]- (2) | 4-cis-O-Caffeoylquinic acid |

**TABLE 7.1** *(Continued)*

| N° | LC Rt (min) | DAD UV bands (nm) | ESI(+)-QToF/MS Exp. Acc. Mass [M+H]⁺ | Error (mDa) | Formula [M+H]⁺ | Adducts & fragment ions of [M+H]⁺ m/z | Exp. Acc. Mass [M−H]⁻ | Error (mDa) | Formula [M−H]⁻ | ESI(−)-QToF/MS Adducts & fragment ions of [M−H]⁻ m/z | Assignment Tentative identification |
|---|---|---|---|---|---|---|---|---|---|---|---|
| *p-Coumaroylquinic acids* | | | | | | | | | | | |
| 8 | 9.82 | 312 | 339.1075 | −0.5 | $C_{16}H_{19}O_8$ | [2M+Na]⁺ 699.188 / [M+Na]⁺ 361.0892 / [pCoumaroyl+H]⁺ 147.0451 / [pCoumaroyl+H-CO]⁺ 119.0500 / [pCoumaroyl+H-2CO]⁺ 91.0556 | 337.0921 | −0.2 | $C_{16}H_{17}O_8$ | [2M−H]⁻ 675.1904 / [Quin−H]⁻ 191.0467 / [pCoumaric−H]⁻ 163.0393 / [pCoumaric−H-CO₂]⁻ 119.0496 | 3-*p*-Coumaroylquinic acid |
| 9 | 13.74 | 308 | 339.1133 | 5.3 | $C_{16}H_{19}O_8$ | [2M+Na]⁺ 699.1916 / [M+Na]⁺ 361.0907 / [pCoumaroyl+H-H₂O]⁺ 147.0453 / [pCoumaroyl+H-H₂O-CO]⁺ 119.0500 / [pCoumaroyl+H-H₂O-2CO]⁺ 91.0561 | 337.0919 | −0.4 | $C_{16}H_{17}O_8$ | [Quin−H]⁻ 191.0553 / [Quin−H-H₂O]⁻ 173.0449 / [pCoumaric−H]⁻ 163.0390 / [pCoumaric−H-CO₂]⁻ 119.0491 | 5-*p*-Coumaroylquinic acid |
| *Caffeoyltartaric acid* | | | | | | | | | | | |
| 10 | 9.06 | 301 sh, 323 | | | | | 311.0526 | −12.3 | $C_{13}H_{11}O_9$ | [Caftar−H-H₂O]⁻ 293.0287 / [Caffeic−H]⁻ 179.0349 / [Tartaric−H]⁻ 149.0227 / [Caffeic−H-CO₂]⁻ 135.0432 | Caffeoyltartaric acid |
| *p-Coumaroyltartaric acid* | | | | | | | | | | | |
| 11 | 15.63 | 310 | | | | | 295.0457 | −0.3 | $C_{13}H_{11}O_8$ | [pCoumaric−H]⁻ 163.0393 / [Tartaric−H]⁻ 149.0104 / [pCoumaroyl+H-CO₂]⁻ 119.0481 | *p*-Coumaroyltartaric acid |
| *Caffeoylmalic acid* | | | | | | | | | | | |
| 12 | 9.05 | 301 sh, | 297.0585 | −2.5 | $C_{13}H_{13}O_8$ | [M+Na]⁺ 319.0429 | 295.0448 | −0.6 | $C_{13}H_{11}O_8$ | [2M−H]⁻ 591.0983 | Caffeoylmalic acid |

**TABLE 7.1**  (*Continued*)

| N° | LC Rt (min) | DAD UV bands (nm) | ESI(+)-QToF/MS Exp. Acc. Mass [M+H]⁺ | Error (mDa) | Formula [M+H]⁺ | Adducts & fragment ions of [M+H]⁺ m/z | ESI(−)-QToF/MS Exp. Acc. Mass [M−H]⁻ | Error (mDa) | Formula [M−H]⁻ | Adducts & fragment ions of [M−H]⁻ m/z | Assignment Tentative identification |
|---|---|---|---|---|---|---|---|---|---|---|---|
| | | 323 | | | | 163.0404 [Caffeoyl+H]⁺; 145.0297 [Caffeoyl+H-H₂O]⁺; 135.0447 [Caffeoyl+H-CO]⁺; 117.0348 [Caffeoyl+H-CO-H₂O]⁺; 89.0397 [Caffeoyl+H-H₂O-2CO]⁺ | | | | 179.0345 [Caffeic-H]⁻; 135.0446 [Caffeic-H-CO₂]⁻; 133.0275 [Malic-H]⁻; 115.0032 [Malic-H-H₂O]⁻; 105.0342 [Malic-H-CO]⁻ | |

*Dicaffeoylquinic acids and caffeoylquinic acid glycosides*

| N° | LC Rt (min) | DAD UV bands (nm) | ESI(+)-QToF/MS Exp. Acc. Mass [M+H]⁺ | Error (mDa) | Formula [M+H]⁺ | Adducts & fragment ions of [M+H]⁺ m/z | ESI(−)-QToF/MS Exp. Acc. Mass [M−H]⁻ | Error (mDa) | Formula [M−H]⁻ | Adducts & fragment ions of [M−H]⁻ m/z | Assignment Tentative identification |
|---|---|---|---|---|---|---|---|---|---|---|---|
| 13 | 5.86 | – | 517.1548 | 0.9 | $C_{22}H_{29}O_{14}$ | 539.1364 [M+Na]⁺; 355.1038 [M-hexosyl]⁺; 163.0415 [Caffeoyl+H]⁺; 145.0310 [Caffeoyl+H-H₂O]⁺; 135.0449 [Caffeoyl+H-CO]⁺; 117.0385 [Caffeoyl+H-CO-H₂O]⁺; 89.0399 [Caffeoyl+H-H₂O-2CO]⁺ | 515.1402 | 0.1 | $C_{22}H_{27}O_{14}$ | 353.0869 [Cafquin-H]⁻; 191.0548 [Quin-H]⁻ | Caffeoylquinic acid-hexoside |
| 14 | 7.56 | – | | | $C_{22}H_{29}O_{14}$ | 539.1367 [M+Na]⁺ | 515.1402 | 0.1 | $C_{22}H_{27}O_{14}$ | | Caffeoylquinic acid-hexoside |
| 15 | 20.20 | 321 | 517.1423 | 7.7 | $C_{25}H_{25}O_{12}$ | 539.1155 [M+Na]⁺; 499.1237 [M+H-H₂O]⁺; 355.0985 [Cafquin+H]⁺; 163.0403 [Caffeoyl+H]⁺; 145.0159 [Caffeoyl+H-H₂O]⁺; 135.0451 [Caffeoyl+H-CO]⁺; 117.0350 [Caffeoyl+H-CO-H₂O]⁺; 89.0404 [Caffeoyl+H-H₂O-2CO]⁺ | 515.1194 | 0.4 | $C_{25}H_{23}O_{12}$ | 353.0871 [Cafquin-H]⁻; 335.0771 [Cafquin-H-H₂O]⁻; 191.0558 [Quin-H]⁻; 179.0349 [Caffeic-H]⁻; 135.0448 [Caffeic-H-CO₂]⁻ | 1,5-di-*O*-Caffeoylquinic acid |
| 16 | 20.63 | 326 | 517.1332 | -1.4 | $C_{25}H_{25}O_{12}$ | 539.1155 [M+Na]⁺ | 515.1186 | -0.4 | $C_{25}H_{23}O_{12}$ | 353.0866 [Cafquin-H]⁻ | 3,5-di-*O*-Caffeoylquinic acid |

**TABLE 7.1** (Continued)

| N° LC | Rt (min) | DAD UV bands (nm) | ESI(+)-QToF/MS Exp. Acc. Mass [M+H]+ | Error (mDa) | Formula [M+H]+ | Adducts & fragment ions of [M+H]+ m/z | ESI(–)-QToF/MS Exp. Acc. Mass [M–H]– | Error (mDa) | Formula [M–H]– | Adducts & fragment ions of [M–H]– m/z | Assignment Tentative identification |
|---|---|---|---|---|---|---|---|---|---|---|---|
| 17 | 24.17 | 331 | 517.1423 | 7.7 | C₂₅H₂₅O₁₂ | 499.1230 [M+H–H₂O]⁺ <br> 355.1016 [Cafquin+H]⁺ <br> 163.0401 [Caffeoyl+H]⁺ <br> 145.0291 [Caffeoyl+H–H₂O]⁺ <br> 135.0450 [Caffeoyl+H–CO]⁺ <br> 117.0346 [Caffeoyl+H–CO–H₂O]⁺ <br> 89.0401 [Caffeoyl+H–H₂O–2CO]⁺ | 515.1190 | 0.0 | C₂₅H₂₃O₁₂ | 335.0761 [Cafquin–H–H₂O]⁺ <br> 191.0556 [Qun–H]⁻ <br> 179.0347 [Caffeic–H]⁻ <br> 135.0446 [Caffeic–H–CO₂]⁻ | 4,5-di-*O*-Caffeoylquinic acid |
| | | | 539.1165 | | C₂₅H₂₅O₁₂ | 499.1228 [M+H–H₂O]⁺ <br> 473.2006 [M+H–CO₂]⁺ <br> 355.0161 [Cafquin+H]⁺ <br> 163.0395 [Caffeoyl+H]⁺ <br> 135.0447 [Caffeoyl+H–CO]⁺ <br> 117.0347 [Caffeoyl+H–CO–H₂O]⁺ <br> 89.0400 [Caffeoyl+H–H₂O–2CO]⁺ | 353.0860 | | | 335.0802 [Cafquin–H–H₂O]⁺ <br> 179.0347 [Caffeic–H]⁻ <br> 173.0449 [Qun–H–H₂O]⁻ <br> 135.0441 [Caffeic–H–CO₂]⁻ <br> [Cafquin–H]⁻ | |

*p-Coumaroylcaffeoylquinic acids*

| N° LC | Rt (min) | DAD UV bands (nm) | ESI(+)-QToF/MS Exp. Acc. Mass [M+H]+ | Error (mDa) | Formula [M+H]+ | Adducts & fragment ions of [M+H]+ m/z | ESI(–)-QToF/MS Exp. Acc. Mass [M–H]– | Error (mDa) | Formula [M–H]– | Adducts & fragment ions of [M–H]– m/z | Assignment Tentative identification |
|---|---|---|---|---|---|---|---|---|---|---|---|
| 18 | 23.58 | 312 | 501.1384 | 1.3 | C₂₅H₂₅O₁₁ | 523.1219 [M+Na]⁺ <br> 483.1295 [M+H–H₂O]⁺ <br> 163.0399 [Caffeoyl+H–H₂O]⁺ <br> 147.0446 [pCoumaroyl+H]⁺ <br> 145.0279 [Caffeoyl+H–2H₂O]⁺ <br> 135.0455 [Caffeoyl+H–H₂O–CO]⁺ <br> 119.0497 [pCoumaroyl+H–CO]⁺ <br> 117.0335 [Caffeoyl+H–2H₂O–CO]⁺ | 499.1233 | 0.7 | C₂₅H₂₃O₁₁ | 353.0868 [M–H–coumaroyl]⁻ <br> 337.0916 [M–H–caffeoyl]⁻ <br> 191.0560 [Qun–H]⁻ <br> 179.0353 [Caffeic–H]⁻ <br> 163.0398 [pCoumaric–H]⁻ <br> 135.0452 [Caffeic–H–CO₂]⁻ <br> 119.0503 [pCoumaric–H–CO₂]⁻ | 3-*p*-Coumaroyl-4-caffeoylquinic acid |

**TABLE 7.1** (Continued)

| Nº | LC Rt (min) | DAD UV bands (nm) | ESI(+)-QToF/MS Exp. Acc. Mass [M+H]+ | Error (mDa) | Formula [M+H]+ | Adducts & fragment ions of [M+H]+ m/z | ESI(−)-QToF/MS Exp. Acc. Mass [M−H]− | Error (mDa) | Formula [M−H]− | Adducts & fragment ions of [M−H]− m/z | Tentative identification |
|---|---|---|---|---|---|---|---|---|---|---|---|
| 19 | 23.95 | 316 | 501.1377 | 2.0 | $C_{23}H_{25}O_{11}$ | 523.1216 [M+Na]+<br>483.1281 [M+H−H₂O]+<br>147.0445 [pCoumaroyl+H]+<br>119.0493 [pCoumaroyl+H−CO]+<br>91.0550 [pCoumaroyl+H−2CO]+<br>91.0550 [pCoumaroyl+H−H₂O−2CO]+<br>89.0398 [Caffeoyl+H−2H₂O−2CO]+ | 499.1241 | −0.1 | $C_{23}H_{23}O_{11}$ | 353.0852 [M−H−coumaroyl]−<br>337.0928 [M−H−caffeoyl]−<br>191.0553 [Quin−H]−<br>179.0342 [Caffeic−H]−<br>163.0390 [pCoumaric−H]−<br>135.0448 [Caffeic−H−CO₂]−<br>119.0490 [pCoumaric−H−CO₂]− | 4-Caffeoyl-5-p-coumaroylquinic acid |

*Dicaffeoyltartaric acids*

| Nº | LC Rt (min) | DAD UV bands (nm) | ESI(+)-QToF/MS Exp. Acc. Mass [M+H]+ | Error (mDa) | Formula [M+H]+ | Adducts & fragment ions of [M+H]+ m/z | ESI(−)-QToF/MS Exp. Acc. Mass [M−H]− | Error (mDa) | Formula [M−H]− | Adducts & fragment ions of [M−H]− m/z | Tentative identification |
|---|---|---|---|---|---|---|---|---|---|---|---|
| 20 | 10.53 | 301 sh, 324 | | | $C_{22}H_{19}O_{12}$ | 497.0677 [M+Na]+<br>457.0698 [M+H−H₂O]+<br>295.0577 [Caftar−H−H₂O]+<br>163.0397 [Caffeoyl+H]+<br>145.0292 [Caffeoyl+H−H₂O]+<br>135.0448 [Caffeoyl+H−CO]+<br>117.0343 [Caffeoyl+H−CO−H₂O]+<br>89.0396 [Caffeoyl+H−H₂O−2CO]+ | 473.0719 | −0.1 | $C_{22}H_{17}O_{12}$ | 947.1354 [2M−H]−<br>311.0402 [Caftar−H]−<br>293.0296 [Caftar−H−H₂O]−<br>179.0345 [Caffeic−H]−<br>149.0091 [Tartaric−H]−<br>135.0443 [Caffeic−H−CO₂]−<br>105.0339 [Tartaric−H−CO₂]− | di-O-Caffeoyltartaric acid |
| 21 | 12.54 | 301 sh, 323 | | | $C_{22}H_{19}O_{12}$ | 295.0563 [Caftar−H−H₂O]+<br>163.0398 [Caffeoyl+H]+<br>145.0288 [Caffeoyl+H−H₂O]+<br>135.0446 [Caffeoyl+H−CO]+<br>117.0341 [Caffeoyl+H−CO−H₂O]+<br>89.0398 [Caffeoyl+H−H₂O−2CO]+ | 473.0713 | −0.7 | $C_{22}H_{17}O_{12}$ | 311.0387 [Caftar−H]−<br>293.0297 [Caftar−H−H₂O]−<br>179.0346 [Caffeic−H]−<br>149.0126 [Tartaric−H]−<br>135.0448 [Caffeic−H−CO₂]−<br>105.0343 [Tartaric−H−CO₂]− | meso-di-O-Caffeoyltartaric acid |

**TABLE 7.1** (*Continued*)

| N° | LC Rt (min) | DAD UV bands (nm) | ESI(+)-QToF/MS | | | | ESI(-)-QToF/MS | | | | Assignment |
|---|---|---|---|---|---|---|---|---|---|---|---|
| | | | Exp. Acc. Mass [M+H]+ | Error (mDa) | Formula [M+H]+ | Adducts & fragment ions of [M+H]+ m/z | Exp. Acc. Mass [M-H]- | Error (mDa) | Formula [M-H]- | Adducts & fragment ions of [M-H]- m/z | Tentative identification |
| *Other hydroxycinnamic acids* | | | | | | | | | | | |
| 22 | 5.39 | – | 343.1098 | 6.9 | C$_{15}$H$_{19}$O$_9$ | 365.0878 [M+Na]+<br>163.0394 [Caffeoyl+H]+<br>145.0104 [Caffeoyl+H-H$_2$O]+<br>135.0497 [Caffeoyl+H-CO]+<br>89.0401 [Caffeoyl+H-H$_2$O-2CO]+ | 341.0905 | -3.2 | C$_{15}$H$_{17}$O$_9$ | | Caffeic acid-hexoside |
| 23 | 5.64 | – | | | C$_{15}$H$_{19}$O$_9$ | 365.0833 [M+Na]+<br>163.0389 [Caffeoyl+H]+<br>145.0289 [Caffeoyl+H-H$_2$O]+<br>135.0473 [Caffeoyl+H-CO]+<br>117.0309 [Caffeoyl+H-CO-H$_2$O]+ | 341.0854 | 1.9 | C$_{15}$H$_{17}$O$_9$ | 179.0330 [Caffeic-H]-<br>135.0435 [Caffeic-H-CO$_2$]- | Caffeic acid-hexoside |
| 24 | 6.08 | 301 sh, 325 | | | C$_{15}$H$_{19}$O$_9$ | 365.0844 [M+Na]+ | 341.0873 | 0.0 | C$_{15}$H$_{17}$O$_9$ | 179.0348 [Caffeic-H]-<br>135.0452 [Caffeic-H-CO$_2$]- | Caffeic acid-hexoside |
| 25 | 7.69 | – | | | C$_{15}$H$_{19}$O$_9$ | 365.0843 [M+Na]+ | 341.0876 | -0.3 | C$_{15}$H$_{17}$O$_9$ | 179.0351 [Caffeic-H]-<br>135.0449 [Caffeic-H-CO$_2$]- | Caffeic acid-hexoside |
| 26 | 8.44 | – | | | C$_{15}$H$_{19}$O$_9$ | 365.0855 [M+Na]+<br>163.0405 [Caffeoyl+H]+<br>145.0137 [Caffeoyl+H-H$_2$O]+<br>135.0455 [Caffeoyl+H-CO]+<br>117.0343 [Caffeoyl+H-CO-H$_2$O]+<br>89.0383 [Caffeoyl+H-H$_2$O-2CO]+ | 341.0867 | 0.6 | C$_{15}$H$_{17}$O$_9$ | 179.0349 [Caffeic-H]-<br>135.0432 [Caffeic-H-CO$_2$]- | Caffeic acid-hexoside |
| 27 | 9.01 | – | | | | | 341.0897 | -2.4 | C$_{15}$H$_{17}$O$_9$ | 179.0349 [Caffeic-H]-<br>135.0432 [Caffeic-H-CO$_2$]- | Caffeic acid-hexoside |
| 28 | 9.52 | – | | | C$_{15}$H$_{19}$O$_9$ | 365.0837 [M+Na]+<br>145.0078 [Caffeoyl+H-H$_2$O]+<br>135.0471 [Caffeoyl+H-CO]+<br>117.0334 [Caffeoyl+H-CO-H$_2$O]+ | 341.0883 | -1.0 | C$_{15}$H$_{17}$O$_9$ | 179.0355 [Caffeic-H]-<br>135.0448 [Caffeic-H-CO$_2$]- | Caffeic acid-hexoside |

**TABLE 7.1** (Continued)

| N° | LC Rt (min) | DAD UV bands (nm) | ESI(+)-QToF/MS | | | | ESI(−)-QToF/MS | | | | Assignment |
|---|---|---|---|---|---|---|---|---|---|---|---|
| | | | Exp. Acc. Mass [M+H]+ | Formula [M+H]+ | Error (mDa) | Adducts & fragment ions of [M+H]+ m/z | Exp. Acc. Mass [M−H]− | Error (mDa) | Formula [M−H]− | Adducts & fragment ions of [M−H]− m/z | Tentative identification |
| 29 | 9.64 | – | | $C_{15}H_{19}O_9$ | | 89.0275 [Caffeoyl+H-H2O-2CO]+; 163.0380 [Caffeoyl+H]+; 145.0338 [Caffeoyl+H-H2O]+; 135.0482 [Caffeoyl+H-CO]+; 117.0348 [Caffeoyl+H-CO-H2O]+; 89.0275 [Caffeoyl+H-H2O-2CO]+ | 341.0897 | −2.4 | $C_{15}H_{17}O_9$ | 135.0442 [Caffeic-H-CO2]− | Caffeic acid-hexoside |
| 30 | 8.01 | 301 sh, 325 | 359.0802 | $C_{18}H_{15}O_8$ | 3.5 | 163.0415 [Caffeoyl+H]+; 145.0640 [Caffeoyl+H-H2O]+; 135.0390 [Caffeoyl+H-CO]+; 117.0346 [Caffeoyl+H-CO-H2O]+; 89.0407 [Caffeoyl+H-H2O-2CO]+ | 357.0633 | −2.3 | $C_{18}H_{13}O_8$ | | Caffeoyl-derivative |
| 31 | 6.03 | 301 sh, 326 | | $C_{17}H_{23}O_{10}$ | | 409.1092 [M+Na]+; 225.0745 [M+H-hexosyl]+ | 385.1138 | −0.3 | $C_{17}H_{21}O_{10}$ | 208.0659 [M-H-hexosyl-CH3]−; 179.0350 [M-H-hexosyl-CO2]−; 164.0519 [M-H-hexosyl-CH3-CO2]−; 149.0620 [M-H-hexosyl-2CH3-CO2]− | Sinapic acid-hexoside |
| 32 | 9.70 | – | | $C_{17}H_{23}O_{10}$ | | 409.0938 [M+Na]+; 225.0774 [M+H-hexosyl]+; 207.0665 [M+H-hexosyl-H2O]+; 192.0411 [M+H-hexosyl-H-CH3OH]+; 175.0411 [M+H-hexosyl-H2O-CH3OH]+; 129.0381 [M+H-hexosyl-2H2O-CO-CH3OH]+ | 385.1117 | 1.8 | $C_{17}H_{21}O_{10}$ | 223.0605 [M-H-hexosyl]−; 208.0372 [M-H-hexosyl-CH3]−; 179.0725 [M-H-hexosyl-CO2]−; 164.0486 [M-H-hexosyl-CH3-CO2]−; 149.0222 [M-H-hexosyl-2CH3-CO2]− | Sinapic acid-hexoside |
| 33 | 10.36 | – | | $C_{17}H_{23}O_{10}$ | | 409.1115 [M+Na]+; 192.0430 [M+H-hexosyl-H-CH3OH]+ | 385.1124 | 1.1 | $C_{17}H_{21}O_{10}$ | 223.0598 [M-H-hexosyl]− | Sinapic acid-hexoside |
| 34 | 13.13 | – | | $C_{17}H_{23}O_{10}$ | | 409.1111 [M+Na]+ | 385.1112 | 2.3 | $C_{17}H_{21}O_{10}$ | 223.0598 [M-H-hexosyl]− | Sinapic acid-hexoside |

**TABLE 7.1** (Continued)

| N° | LC Rt (min) | DAD UV bands (nm) | ESI(+)-QToF/MS Formula [M+H]⁺ | Exp. Acc. Mass [M+H]⁺ | Error (mDa) | Adducts & fragment ions of [M+H]⁺ m/z | Exp. Acc. Mass [M–H]⁻ | Error (mDa) | ESI(–)-QToF/MS Formula [M–H]⁻ | Adducts & fragment ions of [M–H]⁻ m/z | Assignment Tentative identification |
|---|---|---|---|---|---|---|---|---|---|---|---|
| 35 | 8.32 | – | $C_{15}H_{19}O_8$ | | | 225.0753 [M+H-hexosyl]⁺<br>207.0620 [M+H-hexosyl-H₂O]⁺<br>192.0416 [M+H-hexosyl-H-CH₃OH]⁺<br>175.0461 [M+H-hexosyl-H₂O-CH₃OH]⁺<br>129.0322 [M+H-hexosyl-2H₂O-CO-CH₃OH]⁺<br>349.0901 [M+Na]⁺<br>147.0449 [pCoumaroyl+H-H₂O]⁺<br>119.0506 [pCoumaroyl+H-H₂O-CO]⁺<br>91.0569 [pCoumaroyl+H-H₂O-2CO]⁺ | 325.0914 | 0.9 | $C_{15}H_{17}O_8$ | 208.0365 [M-H-hexosyl-CH₃]⁻<br>179.0576 [M-H-hexosyl-CO₃]⁻<br>164.0473 [M-H-hexosyl-CH₃-CO₂]⁻<br>149.0234 [M-H-hexosyl-2CH₃-CO₂]⁻<br>163.0397 [M-H-hexosyl]⁻<br>119.0493 [M-H-hexosyl-CO₂]⁻ | *p*-Coumaric acid-hexoside |
| 36 | 3.70 | – | $C_{15}H_{21}O_9$ | | | 367.0989 [M+Na]⁺ | 343.1029 | 0.0 | $C_{15}H_{19}O_9$ | 181.0496 [DihytroCaf-H]⁻<br>163.0393 [DihytroCaf-H-H₂O]⁻<br>135.0450 [DihytroCaf-H-H₂O-CO]⁻<br>119.0489 [DihytroCaf-H-H₂O-CO₂]⁻ | Dihydrocaffeic acid-hexoside |
| 37 | 3.83 | – | $C_{15}H_{21}O_9$ | | | 367.0999 [M+Na]⁺ | 343.1028 | 0.1 | $C_{15}H_{19}O_9$ | 181.0504 [DihytroCaf-H]⁻<br>163.0398 [DihytroCaf-H-H₂O]⁻<br>135.0450 [DihytroCaf-H-H₂O-CO]⁻<br>119.0492 [DihytroCaf-H-H₂O-CO₂]⁻ | Dihydrocaffeic acid-hexoside |
| 38 | 11.81 | 307 | $C_{11}H_{13}O_4$ | | | | 207.0650 | 0.7 | $C_{11}H_{11}O_4$ | 192.0422 [M-H-CH₃]⁻<br>177.0206 [M-H-2CH₃]⁻ | Ferulic acid methyl ester |
| 39 | 14.47 | – | $C_{11}H_{13}O_4$ | | | | 207.0663 | -0.6 | $C_{11}H_{11}O_4$ | 192.0422 [M-H-CH₃]⁻<br>133.0685 [M-H-CH₃-CO₂]⁻ | Ferulic acid methyl ester |
| 40 | 16.48 | – | $C_{11}H_{13}O_4$ | | | | 207.0656 | 0.1 | $C_{11}H_{11}O_4$ | 192.0435 [M-H-CH₃]⁻<br>177.0206 [M-H-2CH₃]⁻ | Ferulic acid methyl ester |

**TABLE 7.1** (Continued)

| N° | LC Rt (min) | DAD UV bands (nm) | ESI(+)-QToF/MS | | | | ESI(−)-QToF/MS | | | | Assignment Tentative identification |
|---|---|---|---|---|---|---|---|---|---|---|---|
| | | | Exp. Acc. Mass [M+H]⁺ | Error (mDa) | Formula [M+H]⁺ | Adducts & fragment ions of [M+H]⁺ m/z | Exp. Acc. Mass [M−H]⁻ | Error (mDa) | Formula [M−H]⁻ | Adducts & fragment ions of [M−H]⁻ m/z | |
| **Hydroxybenzoic acid derivatives** | | | | | | | | | | $133.0686$   $[M-H-CH_3-CO_2]^-$ | |
| 41 | 4.67 | – | | 3.6 | $C_7H_6O_4$ | $138.0281$   $[M]^{\cdot}$ | $137.0238$ | 0.1 | $C_7H_5O_3$ | $109.0294$   $[M-H-CO]^-$ <br> $93.0331$   $[M-H-CO_2]^-$ | Hydroxybenzoic acid |
| 42 | 5.42 | – | | | $C_7H_6O_4$ | | $153.0196$ | −0.8 | $C_7H_5O_4$ | $135.0448$   $[DHBZ-H-H_2O]^-$ <br> $109.0294$   $[M-H-CO_2]^-$ | Dihydroxybenzoic acid |
| 43 | 4.22 | – | | | $C_{13}H_{17}O_8$ | | $299.0733$ | 3.4 | $C_{13}H_{15}O_8$ | $271.0141$   $[M-H-CO]^-$ <br> $137.0216$   $[HBZ-H]^-$ <br> $93.0498$   $[HBZ-H-CO_2]^-$ | Hydroxybenzoic acid–hexoside |
| 44 | 5.15 | – | | | $C_{13}H_{17}O_8$ | | $299.0764$ | 0.3 | $C_{13}H_{15}O_8$ | $137.0244$   $[HBZ-H]^-$ | Hydroxybenzoic acid–hexoside |
| 45 | 2.49 | – | | | $C_{13}H_{17}O_9$ | | $315.0714$ | 0.2 | $C_{13}H_{15}O_9$ | $153.0181$   $[DHBZ-H]^-$ <br> $152.0114$   $[DHBZ-2H]^-$ <br> $135.0441$   $[DHBZ-H-H_2O]^-$ <br> $109.0283$   $[DHBZ-H-CO_2]^-$ | Dihydroxybenzoic acid–hexoside |
| 46 | 2.69 | – | | | $C_{13}H_{17}O_9$ | | $315.0714$ | 0.2 | $C_{13}H_{15}O_9$ | $153.0181$   $[DHBZ-H]^-$ <br> $152.0114$   $[DHBZ-2H]^-$ <br> $135.0441$   $[DHBZ-H-H_2O]^-$ <br> $109.0283$   $[DHBZ-H-CO_2]^-$ | Dihydroxybenzoic acid–hexoside |
| 47 | 3.74 | – | | | $C_{13}H_{17}O_9$ | | $315.0716$ | 0.0 | $C_{13}H_{15}O_9$ | $153.0185$   $[DHBZ-H]^-$ <br> $109.0287$   $[DHBZ-H-CO_2]^-$ | Dihydroxybenzoic acid–hexoside |
| 48 | 3.91 | – | | | $C_{13}H_{17}O_9$ | | $315.0716$ | 0.0 | $C_{13}H_{15}O_9$ | $153.0172$   $[DHBZ-H]^-$ <br> $109.0307$   $[DHBZ-H-CO_2]^-$ | Dihydroxybenzoic acid–hexoside |

**TABLE 7.1** *(Continued)*

| N° | LC Rt (min) | DAD UV bands (nm) | ESI(+)-QToF/MS Exp. Acc. Mass [M+H]⁺ | Error (mDa) | Formula [M+H]⁺ | Adducts & fragment ions of [M+H]⁺ m/z | ESI(–)-QToF/MS Exp. Acc. Mass [M–H]⁻ | Error (mDa) | Formula [M–H]⁻ | Adducts & fragment ions of [M–H]⁻ m/z | Assignment Tentative identification |
|---|---|---|---|---|---|---|---|---|---|---|---|
| 49 | 4.48 | – | 315.0716 | 0.0 | $C_{13}H_{17}O_9$ | | 315.0716 | 0.0 | $C_{13}H_{15}O_9$ | 153.0172 [DiHBZ-H]⁻ | Dihydroxybenzoic acid-hexoside |
| | | | | | | | | | | 152.0108 [DiHBZ-2H]⁻ | |
| | | | | | | | | | | 135.0441 [DiHBZ-H-H₂O]⁻ | |
| | | | | | | | | | | 109.0261 [DiHBZ-H-CO₂]⁻ | |
| 50 | 4.68 | – | | | $C_{13}H_{17}O_9$ | | 315.0717 | –0.1 | $C_{13}H_{15}O_9$ | 153.0196 [DiHBZ-H]⁻ | Dihydroxybenzoic acid-hexoside |
| | | | | | | | | | | 135.0442 [DiHBZ-H-H₂O]⁻ | |
| | | | | | | | | | | 109.0298 [DiHBZ-H-CO₂]⁻ | |
| 51 | 2.80 | – | | | | | 331.0661 | –0.4 | $C_{13}H_{15}O_{10}$ | 313.0557 [M-H-H₂O]⁻ | Gallic acid-hexoside |
| | | | | | | | | | | 169.0113 [Gallic-H]⁻ | |
| | | | | | | | | | | 168.0057 [Gallic-2H]⁻ | |
| | | | | | | | | | | 149.9953 [Gallic-2H-H₂O]⁻ | |
| | | | | | | | | | | 125.0226 [Gallic-H-CO₂]⁻ | |
| 52 | 2.88 | – | | | | | 331.0661 | –0.4 | $C_{13}H_{15}O_{10}$ | 313.0557 [M-H-H₂O]⁻ | Gallic acid-hexoside |
| | | | | | | | | | | 169.0113 [Gallic-H]⁻ | |
| | | | | | | | | | | 168.0057 [Gallic-2H]⁻ | |
| | | | | | | | | | | 149.9953 [Gallic-2H-H₂O]⁻ | |
| | | | | | | | | | | 125.0226 [Gallic-H-CO₂]⁻ | |
| 53 | 6.61 | – | | | | | 331.0660 | 0.5 | $C_{13}H_{15}O_{10}$ | 313.0544 [M-H-H₂O]⁻ | Gallic acid-hexoside |
| | | | | | | | | | | 169.0140 [Gallic-H]⁻ | |
| | | | | | | | | | | 168.0054 [Gallic-2H]⁻ | |
| | | | | | | | | | | 149.9953 [Gallic-2H-H₂O]⁻ | |
| | | | | | | | | | | 125.0232 [Gallic-H-CO₂]⁻ | |
| 54 | 5.90 | – | 361.1107 | 2.8 | $C_{15}H_{21}O_{10}$ | 97.0288 [M+H-glucosyl-2CH₃-CO-CO₂]⁺ | 359.0975 | 0.3 | $C_{15}H_{19}O_{10}$ | 197.0454 [M-H-glucosyl]⁻ | Syringic acid-hexoside |
| | | | | | | | | | | 182.0210 [M-H-glucosyl-CH₃]⁻ | |
| | | | | | | | | | | 153.0561 [M-H-glucosyl-CO₂]⁻ | |
| | | | | | | | | | | 138.0337 [M-H-glucosyl-CH₃-CO₂]⁻ | |

**TABLE 7.1** *(Continued)*

| N° | LC Rt (min) | DAD UV bands (nm) | ESI(+)-QToF/MS | | | | ESI(-)-QToF/MS | | | | Assignment |
|---|---|---|---|---|---|---|---|---|---|---|---|
| | | | Exp.Acc. Mass [M+H]⁺ | Error (mDa) | Formula [M+H]⁺ | Adducts & fragment ions of [M+H]⁺ m/z | Exp.Acc. Mass [M-H]⁻ | Error (mDa) | Formula [M-H]⁻ | Adducts & fragment ions of [M-H]⁻ m/z | Tentative identification |
| 55 | 17.09 | – | | | $C_{20}H_{21}O_{12}$ | | 451.0880 | -0.3 | $C_{20}H_{19}O_{12}$ | 123.0105 $[M-H-glucosyl-2CH_3-CO_2]^-$ <br> 331.0682 $[M-H]^-$ <br> 313.0558 $[M-H-H_2O]^-$ <br> 168.0060 $[Gallic-2H]^-$ <br> 124.0160 $[Gallic-2H-CO_2]^-$ | Hydroxybenzoyl gallic acid-hexoside |
| 56 | 24.83 | – | | | $C_{20}H_{21}O_{12}$ | | 451.0865 | 1.2 | $C_{20}H_{19}O_{12}$ | 331.0660 $[M-H]^-$ <br> 313.0544 $[M-H-H_2O]^-$ <br> 168.0054 $[Gallic-2H]^-$ <br> 124.0163 $[Gallic-2H-CO_2]^-$ | Hydroxybenzoyl gallic acid-hexoside |
| 57 | 17.68 | – | | | $C_{20}H_{21}O_{11}$ | | 435.0933 | -0.6 | $C_{20}H_{19}O_{11}$ | 315.0722 $[DiHBZhex-H]^-$ or $[M-OC_6H_4CO]^-$ <br> 153.0184 $[DiHBZ-H]^-$ <br> 152.0126 $[DiHBZ-2H]^-$ <br> 137.0258 $[HBZ-H]^-$ <br> 108.0227 $[DiHBZ-2H-CO_2]^-$ <br> 93.0344 $[HBZ-H-CO_2]^-$ | Hydroxybenzoyl-O-dihydroxybenzoic acid-hexoside |
| 58 | 19.41 | – | | | $C_{20}H_{21}O_{11}$ | | 435.0927 | 0.0 | $C_{20}H_{19}O_{11}$ | 315.0710 $[DiHBZhex-H]^-$ or $[M-OC_6H_4CO]^-$ <br> 153.0192 $[DiHBZ-H]^-$ <br> 108.0189 $[DiHBZ-2H-CO_2]^-$ | Hydroxybenzoyl-O-dihydroxybenzoic acid-hexoside |
| 59 | 23.64 | – | | | $C_{20}H_{21}O_{11}$ | | 435.0920 | 0.7 | $C_{20}H_{19}O_{11}$ | | Hydroxybenzoyl-O-dihydroxybenzoic acid-hexoside |
| 60 | 26.88 | 256, 335 sh | | | $C_{20}H_{21}O_{11}$ | | 435.0925 | 0.2 | $C_{20}H_{19}O_{11}$ | 315.0471 $[DiHBZhex-H]^-$ or $[M-OC_6H_4CO]^-$ <br> 297.0611 $[DiHBZhex-H-H_2O]^-$ | Hydroxybenzoyl-O-dihydroxybenzoic acid-hexoside |

**TABLE 7.1** *(Continued)*

| N° | LC Rt (min) | DAD UV bands (nm) | ESI(+)-QToF/MS Exp. Acc. Mass [M+H]+ | Error (mDa) | Formula [M+H]+ | Adducts & fragment ions of [M+H]+ m/z | ESI(-)-QToF/MS Exp. Acc. Mass [M-H]- | Error (mDa) | Formula [M-H]- | Adducts & fragment ions of [M-H]- m/z | Assignment Tentative identification |
|---|---|---|---|---|---|---|---|---|---|---|---|
| 61 | 27.09 | – | | | $C_{20}H_{21}O_{11}$ | | | | | 152.0117 [DiHBZ-2H]- | Hydroxybenzoyl-O-dihydroxybenzoic acid-hexoside |
| | | | | | | | | | | 137.0238 [HBZ-H]- | |
| | | | | | | | | | | 108.0215 [DiHBZ-2H-CO$_2$]- | |
| | | | | | | | | | | 93.0337 [HBZ-H-CO$_2$]- | |
| | | | | | | | 435.0927 | 0.0 | $C_{20}H_{19}O_{11}$ | 315.0715 [DiHBZhex-H]- or [M-OC$_6$H$_4$CO]- | |
| | | | | | | | | | | 297.0609 [DiHBZhex-H-H$_2$O]- | |
| | | | | | | | | | | 153.0195 [DiHBZ-H]- | |
| | | | | | | | | | | 137.0240 [HBZ-H]- | |
| | | | | | | | | | | 108.0215 [DiHBZ-2H-CO$_2$]- | |
| | | | | | | | | | | 93.0341 [HBZ-H-CO$_2$]- | |

**Hydroxyphenylacetic derivatives**

| N° | LC Rt (min) | DAD UV bands (nm) | ESI(+)-QToF/MS Exp. Acc. Mass [M+H]+ | Error (mDa) | Formula [M+H]+ | Adducts & fragment ions of [M+H]+ m/z | ESI(-)-QToF/MS Exp. Acc. Mass [M-H]- | Error (mDa) | Formula [M-H]- | Adducts & fragment ions of [M-H]- m/z | Assignment Tentative identification |
|---|---|---|---|---|---|---|---|---|---|---|---|
| 62 | 5.60 | – | | | $C_8H_9O_3$ | | 151.0392 | 0.3 | $C_8H_7O_3$ | 123.0439 [M-H-CO]- | 4-hydroxyphenylacetic acid |
| | | | | | | | | | | 107.0500 [M-H-CO$_2$]- | |
| 63 | 5.20 | 270, 276 sh | | | $C_{14}H_{19}O_8$ | | 313.0923 | 0.0 | $C_{14}H_{17}O_8$ | 151.0399 [M-H-glucosyl]- | 4-hydroxyphenylacetic acid-hexoside |
| | | | | | | | | | | 123.0447 [M-H-glucosyl-CO]- | |
| | | | | | | | | | | 107.0499 [M-H-glucosyl-CO$_2$]- | |

**Flavonols**

| N° | LC Rt (min) | DAD UV bands (nm) | ESI(+)-QToF/MS Exp. Acc. Mass [M+H]+ | Error (mDa) | Formula [M+H]+ | Adducts & fragment ions of [M+H]+ m/z | ESI(-)-QToF/MS Exp. Acc. Mass [M-H]- | Error (mDa) | Formula [M-H]- | Adducts & fragment ions of [M-H]- m/z | Assignment Tentative identification |
|---|---|---|---|---|---|---|---|---|---|---|---|
| 64 | 17.16 | 279, 344 | 465.1022 | -1.1 | $C_{21}H_{21}O_{12}$ | 487.0832 [M+Na]+ | 463.0874 | -0.3 | $C_{21}H_{19}O_{12}$ | 301.0341 [Y$_0$]- | Quercetin-O-hexoside |
| | | | | | | 303.0501 [Y$_0$]+ | | | | 255.0237 [Y$_0$-CHO-OH]- | |
| | | | | | | 145.0090 [Y$_0$-CHO-OH-4CO]+ | | | | 227.0332 [Y$_0$-2CO-H$_2$O]- | |
| | | | | | | | | | | 151.0027 [$^{1,3}$A]- | |
| | | | | | | | | | | 133.0685 | |
| 65 | 18.03 | 252, 367 | 465.1007 | -2.6 | $C_{21}H_{21}O_{12}$ | 487.0834 [M+Na]+ | 463.0888 | 1.1 | $C_{21}H_{19}O_{12}$ | 301.0356 [Y$_0$]- | Quercetin-O-hexoside |
| | | | | | | 303.0465 [Y$_0$]+ | | | | 255.0310 [Y$_0$-CHO-OH]- | |

**TABLE 7.1** *(Continued)*

| N° | LC Rt (min) | DAD UV bands (nm) | ESI(+)-QToF/MS Exp. Acc. Mass [M+H]⁺ | Error (mDa) | Formula [M+H]⁺ | Adducts & fragment ions of [M+H]⁺ m/z | ESI(−)-QToF/MS Exp. Acc. Mass [M−H]⁻ | Error (mDa) | Formula [M−H]⁻ | Adducts & fragment ions of [M−H]⁻ m/z | Assignment Tentative identification |
|---|---|---|---|---|---|---|---|---|---|---|---|
| 66 | 20.25 | 252, 330 | 465.1032 | −0.1 | $C_{21}H_{21}O_{12}$ | 229.0492 $[Y_0-CHO-OH-CO]^+$<br>153.0186 $[^{13}A]^+$<br>487.0840 $[M+Na]^+$<br>303.0504 $[Y_0]^+$<br>229.0492 $[Y_0-CHO-OH-CO]^+$ | 463.0880 | 0.3 | $C_{21}H_{19}O_{12}$ | 151.0037 $[^{13}A]^-$<br>107.0137 $[^{13}A-2CO]^-[^{12}B-CO]^-$<br>301.0339 $[Y_0]^-$<br>255.0303 $[Y_0-CHO-OH]^-$<br>151.0039 $[^{13}A]^-$ | Quercetin 3-$O$-galactoside |
| 67 | 18.44 | 254, 349 | 479.0826 | 0.0 | $C_{21}H_{19}O_{13}$ | 501.0644 $[M+Na]^+$<br>303.0507 $[Y_0]^+$<br>257.0443 $[Y_0-CHO-OH]^+$<br>153.0186 $[^{13}A]^+$ | 477.0675 | 1.1 | $C_{21}H_{17}O_{13}$ | 301.0347 $[Y_0]^-$<br>255.0293 $[Y_0-CHO-OH]^-$<br>227.0346 $[Y_0-2CO-H_2O]^-$<br>151.0036 $[^{13}A]^-$ | Quercetin-3-$O$-glucuronide |
| 68 | 9.50 | 256, 352 | 641.1385 | 3.1 | $C_{27}H_{29}O_{18}$ | 663.1232 $[M+Na]^+$<br>303.0515 $[Y_0]^+$ | 639.1168 | −2.9 | $C_{27}H_{27}O_{18}$ | 463.0865 $[Y_1]^-$<br>301.0360 $[Y_0]^-$<br>135.0432 $[^{12}A-CO]^-[^{12}B]^-$ | Quercetin hexose–glucuronide |
| 69 | 10.58 | – | 641.1385 | 3.1 | $C_{27}H_{29}O_{18}$ | 663.1232 $[M+Na]^+$<br>465.1066 $[Y_1]^+$<br>303.0515 $[Y_0]^+$ | 639.1168 | −2.9 | $C_{27}H_{27}O_{18}$ | 463.0865 $[Y_1]^-$<br>301.0360 $[Y_0]^-$ | Quercetin hexose–glucuronide |
| 70 | 21.52 | 255, 352 | 551.1039 | 0.2 | $C_{24}H_{23}O_{15}$ | 573.0847 $[M+Na]^+$<br>303.0508 $[Y_0]^+$<br>273.0406<br>229.0497 $[Y_0-CHO-OH-CO]^+$<br>153.0186 $[^{13}A]^+$<br>145.0516 $[Y_0-CHO-OH-4CO]^+$ | 549.0879 | −0.1 | $C_{24}H_{21}O_{15}$ | 1099.1829 $[2M-H]^-$<br>505.0987 $[M-H-CO_2]^-$<br>463.0865 $[M-H-CO_2-C2H_2O]^-$<br>301.0340 $[Y_0]^-$<br>300.0273 $[Y_0-H]^-$<br>271.0243<br>255.0305 $[Y_0-CHO-OH]^-$<br>151.0038 $[^{13}A]^-$ | Quercetin-3-$O$-malonylglucoside |
| 71 | 22.03 | 252, | 551.1031 | −0.6 | $C_{24}H_{23}O_{15}$ | 573.0846 $[M+Na]^+$ | 549.0891 | 1.1 | $C_{24}H_{21}O_{15}$ | 505.0990 $[M-H-CO_2]^-$ | Quercetin-3-$O$-malonylglucoside |

**TABLE 7.1** *(Continued)*

| Nº | LC Rt (min) | DAD UV bands (nm) | ESI(+)-QToF/MS Exp. Acc. Mass [M+H]+ | Error (mDa) | Formula [M+H]+ | Adducts & fragment ions of [M+H]+ m/z | ESI(−)-QToF/MS Exp. Acc. Mass [M−H]− | Error (mDa) | Formula [M−H]− | Adducts & fragment ions of [M−H]− m/z | Assignment Tentative identification |
|---|---|---|---|---|---|---|---|---|---|---|---|
| | | 364 | | | | [Y0]+ 303.0506; 273.0407; [Y0−CHO−OH−CO]+ 229.0504; [13A]+ 153.0196; [Y0−CHO−OH−4CO]+ 145.0495 | | | | [M−H−CO2−C2H2O]− 463.0880; [Y0]− 301.0351; 271.0244; [Y0−CHO−OH]− 255.0284; [13A]− 151.0033; [02A−2CO]−[02B−CO]− 107.0130 | |
| 72 | 23.69 | – | 551.1041 | 0.4 | $C_{24}H_{23}O_{15}$ | [M+Na]+ 573.0851; [Y0]+ 303.0504; 273.0768; [Y0−CHO−OH−CO]+ 229.0488; [13A]+ 153.0195; 147.0456 | 549.0894 | 1.4 | $C_{24}H_{21}O_{15}$ | [M−H−CO2]− 505.0980; [Y0]− 301.0335; [Y0−H]− 300.0266; 271.0236; [Y0−CHO−OH]− 255.0290; [13A]− 151.0039; [02A−2CO]−[02B−CO]− 107.0127 | Quercetin-3-O-malonylglucoside |
| 73 | 11.51 | 253, 355 | 727.1348 | −1.0 | $C_{30}H_{31}O_{21}$ | [M+Na]+ 749.1142; [Y1]+ 479.0830; [Y0]+ 303.0494 | 725.1176 | −2.5 | $C_{30}H_{29}O_{21}$ | [M−H−CO2]− 681.1274; [M−H−CO2−glucuronyl]− 505.0977; [Y0]− 301.0355; [Y0−CHO−OH]− 255.0300 | Quercetin-3-O-(6″-O-malonyl)-glucoside-7-O-glucuronide |
| 74 | 13.82 | 253, 350 | 713.1565 | 0.0 | $C_{30}H_{33}O_{20}$ | [M+Na]+ 735.1379; [Y1]+ 465.1039; [Y0]+ 303.0508 | 711.1411 | 0.2 | $C_{30}H_{31}O_{21}$ | [M−H−CO2]− 667.1519; [M−H−CO2−hexosyl−C2H2O]− 463.0863; [Y1]− 301.0348; [02A−CO]−[02B]− 135.0641 | Quercetin-3-O-(6″-O-malonyl)-glucoside-7-O-glucoside |
| 75 | 12.18 | – | 627.1580 | 1.9 | $C_{27}H_{31}O_{17}$ | [M+Na]+ 649.1414; [Y0]+ 303.0502 | 625.1391 | −1.4 | $C_{27}H_{29}O_{17}$ | [Y1]− 463.0874; [Y0]− 301.0344 | Quercetin-O-di-hexoside |

**TABLE 7.1**　(Continued)

| N° LC | Rt (min) | DAD UV bands (nm) | ESI(+)-QToF/MS Exp. Acc. Mass [M+H]+ | Error (mDa) | Formula [M+H]+ | Adducts & fragment ions of [M+H]+ m/z | ESI(−)-QToF/MS Exp. Acc. Mass [M−H]− | Error (mDa) | Formula [M−H]− | Adducts & fragment ions of [M−H]− m/z | Assignment Tentative identification |
|---|---|---|---|---|---|---|---|---|---|---|---|
| 76 | 16.07 | – | 627.1556 | −0.5 | C27H31O17 | 137.0611 [0,2A-CO]+ <br> 649.1367 [M+Na]+ <br> 449.1805 [Y1]+ <br> 303.0522 [Y0]+ | 625.1400 | −0.5 | C27H29O17 | 447.0833 [Y1]− <br> 301.0290 [Y0]− | Quercetin-O-rhamnosyl-glucunide |
| 77 | 25.27 | 265, 347 | 535.1094 | 0.6 | C24H23O14 | 557.0905 [M+Na]+ <br> 287.0560 [Y0]+ <br> 121.0301 [0,2B]+ <br> 153.0204 [1,3A]+ | 533.0889 | −3.9 | C24H21O14 | 489.1039 [M-H-CO2]− <br> 285.0399 [Y0]− <br> 255.0298 [Y0-CO-2H]− <br> 227.0343 [Y0-CHO-CO-H]− <br> 151.0037 [1,3A]− <br> 107.0154 [0,2A-2CO]−;[0,2B-CO]− | Kaempferol-3-O-(6″-O-malonyl)-glucoside |
| 78 | 23.90 | – | 449.1092 | 0.8 | C21H21O11 | 471.0901 [M+Na]+ <br> 287.0561 [Y0]+ | 447.0925 | 0.2 | C21H19O11 | 285.0410 [Y0]− <br> 151.0056 | Kaempferol-3-O-glucoside |
| 79 | 26.43 | – | 449.1084 | 0.0 | C21H21O11 | 471.0830 [M+Na]+ <br> 287.0549 [Y0]+ | 447.0925 | −0.1 | C21H19O11 | 285.0406 [Y0]− | Kaempferol-hexoside |
| 80 | 22.34 | 265, 332 | 463.0878 | 0.1 | C21H19O12 | 485.0683 [M+Na]+ <br> 287.0559 [Y0]+ <br> 133.1025 [1,3B-2H]+ | 461.0724 | 0.4 | C21H17O12 | 285.0403 [Y0]− <br> 257.0471 [Y0-CO]− <br> 229.0509 [Y0-2CO]− | Kaempferol-3-O-glucuronide |
| 81 | 27.08 | – | 287.0560 | 0.4 | C15H11O6 | 259.1070 [Y0-CO]+ <br> 213.0885 [Y0-H2O-2CO]+ <br> 185.0970 [Y0-H2O-3CO]+ <br> 171.0856 [Y0-CHO-OH-CO-C2H2O]+ <br> 153.0146 [1,3A]+ <br> 137.0894 [0,2A-CO]+;[0,2B]+ <br> 135.0776 [1,3B-2H]+ | 285.0399 | 0.0 | C15H9O6 | 153.0197 [1,3A]− <br> 137.0239 [0,2A-CO]−;[0,2B]− <br> 133.0310 [1,3B-2H]− <br> 109.0296 [0,2A-2CO]−;[0,2B-CO]− <br> 93.0340 [1,3B-CO]− | Kaempferol |

**TABLE 7.1** (Continued)

| N° | LC Rt (min) | DAD UV bands (nm) | ESI(+)-QToF/MS | | | | ESI(-)-QToF/MS | | | | Assignment |
| --- | --- | --- | --- | --- | --- | --- | --- | --- | --- | --- | --- |
| | | | Exp. Acc. Mass [M+H]⁺ | Error (mDa) | Formula [M+H]⁺ | Adducts & fragment ions of [M+H]⁺ m/z | Exp. Acc. Mass [M-H]⁻ | Error (mDa) | Formula [M-H]⁻ | Adducts & fragment ions of [M-H]⁻ m/z | Tentative identification |
| **Flavones** | | | | | | | | | | | |
| 82 | 19.82 | 255, 347 | 449.1081 | −0.3 | $C_{21}H_{21}O_{11}$ | 471.0901 [M+Na]⁺; 371.1316; 287.0559 $[Y_0]^+$; 153.0177 [1,3A]⁺; 135.0821 [1,3B]⁺; 127.0807 $[Y_0$-CHO-OH-3CO-CH$_2$O]⁺; 121.0653 [1,3B]⁺; 107.0500 [1,3A-H$_2$O-CO]⁺, [1,3B-CO]⁺; 105.0681 [1,3B-2H-CO]⁺ | 447.0925 | −0.2 | $C_{21}H_{19}O_{11}$ | 895.1951 [2M-H]⁻; 285.0400 $[Y_0]^-$; 217.0505 $[Y_0$-C2H$_2$O-C2H$_2$]⁻; 199.0396 $[Y_0$-CHO-2CO-H]⁻; 175.0402 | Luteolin-7-*O*-glucoside |
| 83 | 17.45 | 253, 348 | 463.0880 | 0.3 | $C_{21}H_{19}O_{12}$ | 485.0690 [M+Na]⁺; 287.0559 $[Y_0]^+$; 153.0186 [1,3A]⁺ | 461.0717 | −0.3 | $C_{21}H_{17}O_{12}$ | 923.1496 [2M-H]⁻; 285.0398 $[Y_0]^-$; 217.0506 $[Y_0$-C2H$_2$O-C2H$_2$]⁻; 199.0390 $[Y_0$-CHO-2CO-H]⁻; 175.0358; 151.0032 [1,3A]⁻; 133.0287 [1,3B]⁻; 133.0287 | Luteolin 7-*O*-glucuronide |
| 84 | 20.27 | – | 595.1651 | −1.2 | $C_{27}H_{31}O_{15}$ | | 593.1498 | −0.8 | $C_{27}H_{29}O_{15}$ | 285.0685 $[Y_0]^-$ | Luteolin-7-*O*-rhamnosyl]-hexoside |
| 85 | 21.17 | 268, 351 | 595.1672 | 0.9 | $C_{27}H_{31}O_{15}$ | 617.1484 [M+Na]⁺; 449.1083 $[Y_1]^+$; 371.1316; 287.0557 $[Y_0]^+$ | 593.1498 | −0.8 | $C_{27}H_{29}O_{15}$ | 447.0604 $[Y_1]^-$; 285.0400 $[Y_0]^-$ | Luteolin-7-*O*-rutinoside |
| 86 | 11.85 | 265 | 553.1192 | −0.1 | $C_{24}H_{25}O_{15}$ | 1069.2104 [2(M-H$_2$O)+H]⁺ | 551.1027 | −1.0 | $C_{24}H_{23}O_{15}$ | 1103.2147 [2M-H]⁻ | Luteolin-hydroxymalonylhexoside |

*High-Resolution Mass Spectroscopy for Phytochemical Analysis*

**TABLE 7.1** *(Continued)*

| N° | LC Rt (min) | DAD UV bands (nm) | ESI(+)-QToF/MS Exp. Acc. Mass [M+H]⁺ | Error (mDa) | Formula [M+H]⁺ | Adducts & fragment ions of [M+H]⁺ m/z | ESI(−)-QToF/MS Exp. Acc. Mass [M−H]⁻ | Error (mDa) | Formula [M−H]⁻ | Adducts & fragment ions of [M−H]⁻ m/z | Assignment Tentative identification |
|---|---|---|---|---|---|---|---|---|---|---|---|
| | | 339 | | | | 575.1013 $[M+Na]^+$; 535.1091 $[M+H-H_2O]^+$; 449.0679 $[M-OH-malonyl+H]^+$; 287.0556 $[Y_0]^+$; 137.0244 $[^{0,2}A-CO]^+:[^{0,2}B]^+$ | | | | 507.1135 $[M-H-CO_2]^-$; 371.0977 $[M-H-hex]^-$; 285.0402 $[Y_0]^-$; 241.0501 $[Y_0-CO2]^-$; 255.0302 $[Y_0-CO-2H]^-$; 227.0341 $[Y_0-CHO-CO-H]^-$; 199.0396 $[Y_0-CHO-2CO-H]^-$; 175.0397 | |
| 87 | 20.57 | – | 447.0912 | 1.5 | $C_{21}H_{19}O_{11}$ | 271.0608 $[Y_0]^+$ | 445.0763 | 0.8 | $C_{21}H_{17}O_{11}$ | 269.0449 $[Y_0]^-$ | Apigenin–glucuronide |
| 88 | 23.02 | 259, 328 | 433.1137 | −0.2 | $C_{21}H_{21}O_{10}$ | 271.0610 $[Y_0]^+$ | 431.0972 | 0.6 | $C_{21}H_{19}O_{10}$ | 269.0441 $[Y_0]^-$ | Apigenin–glucoside |
| 89 | 23.90 | – | 579.1711 | 0.3 | $C_{27}H_{31}O_{10}$ | 433.1124 $[Y_1]^+$ | 577.1553 | 0.4 | $C_{27}H_{29}O_{14}$ | 433.2084 $[Y_1]^-$ | Apigenin-*O*-rhamnosyl-hexoside |
| 90 | 14.92 | – | 465.1038 | 0.5 | $C_{21}H_{21}O_{12}$ | 271.0605 $[Y_0]^+$; 487.0856 $[M+Na]^+$; 271.0606 $[Y_0]^+$; 163.0387 $[^{0,4}B]^+$; 153.0191 $[^{1,3}A]^+$; 145.0296 $[^{0,4}B-H_2O]^+$; 121.0292 $[^{0,2}B]^+$; 91.0529 $[^{1,3}B]^+$ | 463.0872 | −0.5 | $C_{21}H_{19}O_{12}$ | 269.0446 $[Y_0]^-$; 269.0446 $[Y_0]^-$ | Apigenin–pentahydroxyhexanoide |
| 91 | 26.99 | – | 839.3358 | −2.0 | $C_{40}H_{55}O_{19}$ | 271.0610 $[Y_0]^+$; 259.1070 $[Y_0-CO]^+$; 213.0885 $[Y_0-H_2O-2CO]^+$; 185.0970 $[Y_0-H_2O-3CO]^+$; 179.0649 $[^{0,4}B]^+$; 153.0146 $[^{1,3}A]^+$; 137.0894 $[^{0,2}A-CO]^+:[^{0,2}B]^+$ | 837.3194 | −1.3 | $C_{40}H_{55}O_{19}$ | 269.0450 $[Y_0]^-$ | Apigenin conjugate |
| 92 | 27.08 | – | 287.0560 | 0.4 | $C_{15}H_{11}O_{6}$ | | 285.0399 | 0.0 | $C_{15}H_{9}O_{6}$ | 153.0197 $[^{1,3}A]^-$; 137.0239 $[^{0,2}A-CO]^-:[^{0,2}B]^-$ | Luteolin |

**TABLE 7.1**  *(Continued)*

| N° | LC Rt (min) | DAD UV bands (nm) | ESI(+)-QToF/MS | | | | ESI(−)-QToF/MS | | | | Assignment |
|---|---|---|---|---|---|---|---|---|---|---|---|
| | | | Exp. Acc. Mass [M+H]$^+$ | Error (mDa) | Formula [M+H]$^+$ | Adducts & fragment ions of [M+H]$^+$ m/z | Exp. Acc. Mass [M−H]$^-$ | Error (mDa) | Formula [M−H]$^-$ | Adducts & fragment ions of [M−H]$^-$ m/z | Tentative identification |
| **Flavanones** | | | | | | | | | | | |
| 93 | 14.87 | 284, 329 sh | 465.1026 | −0.7 | $C_{21}H_{21}O_{12}$ | 135.0776 $[^{1,3}B-2H]^+$<br>117.0767 $[^{1,3}B-H_2O]^+$<br>107.0500 $[^{1,3}A-H_2O-CO]^+$, $[^{1,3}B-CO]^+$ | 463.0882 | 0.5 | $C_{21}H_{19}O_{12}$ | 287.0555 $[Y_0]^-$<br>151.0037 $[^{1,3}A]^-$<br>135.0452 $[^{1,3}B]^-$<br>107.0133 $[^{1,3}A]^-$ | Eriodictyol-*O*-glucuronide |
| **Anthocyanidin** | | | | | | | | | | | |
| 94 | 10.80 | | 449.1079 | 0.5 | $C_{21}H_{20}O_{11}$ | 487.0830 $[M+Na]^+$<br>289.0715 $[Y_0]^+$<br>153.0187 $[^{1,3}A]^+$<br>287.0558 $[Y_0]^+$<br>213.0555 $[Y_0-H_2O-2CO]^+$<br>185.0208 $[Y_0-H_2O-3CO]^+$<br>137.0251 $[^{1,3}B]^+$<br>121.0298 $[^{1,3}A]^+$<br>109.0309 $[^{0,2}B-CO]^+$ | - | - | $C_{21}H_{21}O_{11}$ | | Cyanidin 3-*O*-glucoside |
| 95 | 13.62 | | 535.1089 | −0.1 | $C_{24}H_{23}O_{14}$ | 287.0556 $[Y_0]^+$<br>137.0236 $[^{1,3}B]^+$<br>109.0292 $[^{1,3}B-CO]^+$ | - | - | $C_{24}H_{23}O_{14}$ | 507.1121 $[Y_1-CO]^-$<br>489.1013 $[Y_1-CO-H_2O]^-$<br>285.0390 $[Y_0-2H]^-$ | Cyanidin 3-*O*-(3''-*O*-malonyl)-glucoside |
| 96 | 16.84 | 279, >500 | 535.1095 | −0.7 | $C_{24}H_{23}O_{14}$ | 449.1184 $[M-CH_2CO]^+$<br>287.0561 $[Y_0]^+$<br>213.0555 $[M-H_2O-2CO]^+$<br>137.0246 $[^{1,3}B]^+$<br>109.0293 $[^{1,3}B-CO]^+$ | - | - | $C_{24}H_{23}O_{14}$ | 507.1138 $[M-CO]^-$<br>489.1035 $[M-CO-H_2O]^-$<br>285.0400 $[Y_0-2H]^-$ | Cyanidin 3-*O*-(6''-*O*-malonyl)-glucoside |

**TABLE 7.1** *(Continued)*

| N° | LC Rt (min) | DAD UV bands (nm) | ESI(+)-QToF/MS Exp. Acc. Mass [M+H]+ | Error (mDa) | Formula [M+H]+ | Adducts & fragment ions of [M+H]+ m/z | ESI(-)-QToF/MS Exp. Acc. Mass [M-H]- | Error (mDa) | Formula [M-H]- | Adducts & fragment ions of [M-H]- m/z | Assignment Tentative identification |
|---|---|---|---|---|---|---|---|---|---|---|---|
| 97 | 20.25 | | 491.1166 | 2.4 | $C_{23}H_{23}O_{12}$ | 449.1080 [M-CH$_2$CO]+<br>287.0559 [Y$_0$]+<br>269.0442 [Y$_0$-H$_2$O]+<br>259.0596 [Y$_0$-CO]+ | - | | $C_{23}H_{21}O_{12}$ | | Cyanidin 3-O-(6''-O-acetyl)-glucoside |
| **Coumarins** | | | | | | | | | | | |
| 98 | 6.50 | 290, 340 | 341.0866 | -0.7 | $C_{15}H_{17}O_9$ | 363.0684 [M+Na]+<br>179.0345 [Y$_0$]+<br>133.0284 [Y$_0$-CO-H$_2$O]+<br>123.0456 [Y$_0$-2CO]+ | 339.0727 | 1.1 | $C_{15}H_{15}O_9$ | 399.1273 [M-H+AcO]-<br>177.0188 [Y$_0$]-<br>133.0288 [Y$_0$-CO$_2$]-<br>105.0036 [Y$_0$-CO$_2$-CO]- | Esculetin-6-O-glucoside |
| 99 | 7.31 | – | 179.0341 | 0.3 | $C_9H_7O_4$ | 133.0292 [M+H-CO-H$_2$O]+<br>123.0437 [M+H-2CO]+ | 177.0191 | -0.3 | $C_9H_5O_4$ | 149.0236 [Y$_0$-CO]-<br>133.0288 [Y$_0$-CO$_2$]-<br>105.0341 [Y$_0$-CO$_2$-CO]- | Dihydroxycoumarin |
| 100 | 10.23 | – | 179.0344 | 0.0 | $C_9H_7O_4$ | 133.0289 [M+H-CO-H$_2$O]+<br>123.0452 [M+H-2CO]+ | 177.0192 | -0.4 | $C_9H_5O_4$ | 149.0222 [Y$_0$-CO]-<br>133.0292 [Y$_0$-CO$_2$]-<br>105.0344 [Y$_0$-CO$_2$-CO]- | Dihydroxycoumarin |
| 101 | 12.02 | 296, 330 | 179.0339 | 0.0 | $C_9H_7O_4$ | 133.0288 [M+H-CO-H$_2$O]+<br>123.0421 [M+H-2CO]+ | 177.0187 | 0.1 | $C_9H_5O_4$ | 133.0236 [Y$_0$-CO$_2$]-<br>105.0340 [Y$_0$-CO$_2$-CO]- | 6,7-dihydroxycoumarin |
| 102 | 9.05 | – | 295.0518 | -6.4 | $C_{13}H_{11}O_8$ | 317.0241 [M+Na]+<br>179.0376 [Y$_0$]+<br>133.0286 [Y$_0$-CO-H$_2$O]+<br>123.0463 [Y$_0$-2CO]+ | 293.0295 | 0.2 | $C_{13}H_9O_8$ | 177.0194 [Y$_0$]-<br>149.0243 [Y$_0$-CO]-<br>133.0284 [Y$_0$-CO$_2$]-<br>105.0342 [Y$_0$-CO$_2$-CO]- | Maloyl-dihydroxycoumarin |
| 103 | 10.54 | – | 295.0510 | -5.6 | $C_{13}H_{11}O_8$ | 133.0288 [Y$_0$-CO-H$_2$O]+ | 293.0296 | 0.1 | $C_{13}H_9O_8$ | 177.0187 [Y$_0$]-<br>149.0090 [Y$_0$-CO]-<br>133.0286 [Y$_0$-CO$_2$]-<br>105.0339 [Y$_0$-CO$_2$-CO]- | Maloyl-dihydroxycoumarin |

**TABLE 7.1** (Continued)

| Nº LC | Rt (min) | DAD UV bands (nm) | ESI(+)-QToF/MS Exp. Acc. Mass [M+H]+ | Error (mDa) | Formula [M+H]+ | Adducts & fragment ions of [M+H]+ m/z | ESI(-)-QToF/MS Exp. Acc. Mass [M-H]- | Error (mDa) | Formula [M-H]- | Adducts & fragment ions of [M-H]- m/z | Assignment Tentative identification |
|---|---|---|---|---|---|---|---|---|---|---|---|
| 104 | 12.54 | – | 295.0541 | –8.7 | $C_{13}H_{11}O_8$ | 179.0348 [Y$_0$]+; 133.0446 [Y$_0$-CO-H$_2$O]+ | 293.0299 | –0.2 | $C_{13}H_9O_8$ | 177.0189 [Y$_0$]-; 149.0139 [Y$_0$-CO]-; 133.0290 [Y$_0$-CO$_2$]-; 105.0343 [Y$_0$-CO$_2$-CO]- | Maloyl-dihydroxycoumarin |
| **Hydrolysable tannins** | | | | | | | | | | | |
| 105 | 27.09 | – | | | $C_{30}H_{31}O_{12}$ | | 581.1663 | –0.4 | $C_{30}H_{29}O_{12}$ | 295.0826 [4-hydroxyphenylacetichex-H-H$_2$O]-; 175.0391 [4-hydroxyphenylacetichex-H-H$_2$O-C$_6$H$_5$CH$_3$CO]-; 151.0392 [4-hydroxyphenylacetic-H]-; 143.0344 [4-hydroxyphenylacetichex-H-H$_2$O-C$_6$H$_5$CH$_2$OHCO$_2$]- | Tri-4-hydroxyphenylacetic acid-glucoside |
| **Lignan derivatives** | | | | | | | | | | | |
| 106 | 21.00 | – | | | $C_{22}H_{27}O_8$ | | 417.1569 | –2.0 | $C_{22}H_{25}O_8$ | 359.1021 [M-H-2CH$_2$-CO]- | Syringaresinol |
| 107 | 13.90 | – | | | $C_{28}H_{37}O_{13}$ | 603.2055 [M+Na]+; 383.1479 [M+H-hexosyl-2H$_2$O]+ | 579.2075 | 0.3 | $C_{28}H_{35}O_{13}$ | 417.1544 [M-H-hexosyl]-; 399.1437 [M-H-hexosyl-H$_2$O]- | Syringaresinol-hexose |
| 108 | 18.97 | – | | | $C_{28}H_{37}O_{13}$ | | 579.2104 | –2.6 | $C_{28}H_{35}O_{13}$ | 417.1558 [M-H-hexosyl]- | Syringaresinol-hexose |
| 109 | 19.63 | – | | | $C_{28}H_{37}O_{13}$ | 603.2061 [M+Na]+ | 579.2079 | –0.1 | $C_{28}H_{35}O_{13}$ | 417.1558 [M-H-hexosyl]-; 399.1493 [M-H-hexosyl-H$_2$O]- | Syringaresinol-hexose |
| 110 | 23.30 | – | | | $C_{28}H_{37}O_{13}$ | 603.2059 [M+Na]+; 383.1505 [M+H-hexosyl-2H$_2$O]+ | 579.2075 | 0.3 | $C_{28}H_{35}O_{13}$ | 417.1555 [M-H-hexosyl]-; 387.1104 [M-H-hexosyl-2CH$_3$]- | Syringaresinol-hexose |
| 111 | 15.06 | 205, 280 | | | $C_{30}H_{39}O_{14}$ | | 621.2198 | –1.5 | $C_{30}H_{37}O_{14}$ | 417.1559 [M-H-acetylhexosyl]-; 402.1313 [M-H-acetylhexosyl-CH$_3$]-; 399.1447 [M-H-H$_2$O]- | Syringaresinol-acetylhexose |

**TABLE 7.1** (Continued)

| Nº LC | Rt (min) | DAD UV bands (nm) | ESI(+)-QToF/MS Exp. Acc. Mass [M+H]⁺ | Error (mDa) | Formula [M+H]⁺ | Adducts & fragment ions of [M+H]⁺ m/z | ESI(−)-QToF/MS Exp. Acc. Mass [M−H]⁻ | Error (mDa) | Formula [M−H]⁻ | Adducts & fragment ions of [M−H]⁻ m/z | Adducts & fragment ions of [M−H]⁻ | Assignment Tentative identification |
|---|---|---|---|---|---|---|---|---|---|---|---|---|
| 112 | 24.50 | – | | | C₃₀H₃₉O₁₄ | | 621.2183 | 0.0 | C₃₀H₃₇O₁₄ | 387.1058 | [M-H-2CH₃]⁻ | Syringaresinol-acetylhexose |
| | | | | | | | | | | 417.1548 | [M-H-acetylhexosyl]⁻ | |
| | | | | | | | | | | 402.1313 | [M-H-acetylhexosyl-CH₃]⁻ | |
| | | | | | | | | | | 387.1078 | [M-H-acetylhexosyl-2CH₃]⁻ | |
| | | | | | | | | | | 359.1111 | [M-H-acetylhexosyl-2CH₃-CO]⁻ | |
| | | | | | | | | | | 181.0503 | [M-H-acetylhexosyl-2CH₂O-OH-C6H₂-CHO-2(CH₂CH)]⁻ | |
| | | | | | | | | | | 166.0268 | [M-H-acetylhexosyl-2CH₂O-OH-C6H₂-CHO-2(CH₂CH)-CH₃]⁻ | |
| | | | | | | | | | | 151.0044 | [M-H-acetylhexosyl-2CH₃O-OH-C6H₂-CHO-2(CH₂CH)-2CH₃]⁻ | |
| | | | | | | | | | | 123.0065 | [M-H-acetylhexosyl-2CH₃O-OH-C6H₂-CHO-2(CH₂CH)-2CH₃-CO]⁻ | |
| 113 | 24.63 | – | | | C₃₀H₃₉O₁₄ | | 621.2181 | 0.2 | C₃₀H₃₇O₁₄ | 417.1546 | [M-H-acetylhexosyl]⁻ | Syringaresinol-acetylhexose |
| | | | | | | | | | | 402.1313 | [M-H-acetylhexosyl-CH₃]⁻ | |
| | | | | | | | | | | 387.1074 | [M-H-acetylhexosyl-2CH₃]⁻ | |
| | | | | | | | | | | 359.1084 | [M-H-acetylhexosyl-2CH₃-CO]⁻ | |
| | | | | | | | | | | 181.0503 | [M-H-acetylhexosyl-2CH₂O-OH-C6H₂-CHO-2(CH₂CH)]⁻ | |
| | | | | | | | | | | 166.0269 | [M-H-acetylhexosyl-2CH₂O-OH-C6H₂-CHO-2(CH₂CH)-CH₃]⁻ | |
| | | | | | | | | | | 151.0041 | [M-H-acetylhexosyl-2CH₃O-OH-C6H₂-CHO-2(CH₂CH)-2CH₃]⁻ | |

**TABLE 7.1** *(Continued)*

| N° LC | Rt (min) | DAD UV bands (nm) | ESI(+)-QToF/MS | | | | ESI(−)-QToF/MS | | | | Assignment |
|---|---|---|---|---|---|---|---|---|---|---|---|
| | | | Exp. Acc. Mass $[M+H]^+$ | Error (mDa) | Formula $[M+H]^+$ | Adducts & fragment ions of $[M+H]^+$ $m/z$ | Exp. Acc. Mass $[M-H]^-$ | Error (mDa) | Formula $[M-H]^-$ | Adducts & fragment ions of $[M-H]^-$ $m/z$ | Tentative identification |
| 114 | 19.22 | – | | | $C_{28}H_{39}O_{13}$ | | 581.2239 | −0.5 | $C_{28}H_{37}O_{13}$ | 341.1392 $[M-H-hexosyl-CH_3COOH-H_2O]^-$ <br> 329.1390 $[M-H-hexosyl-CH_3COOH-2CH_3]^-$ | Dimethoxy-hexosyl-lariciresinol |
| 115 | 19.39 | – | | | $C_{28}H_{39}O_{13}$ | | 581.2238 | −0.4 | $C_{28}H_{37}O_{13}$ | 359.1494 $[M-H-hexosyl-CH_3COOH]^-$ <br> 341.1383 $[M-H-hexosyl-CH_3COOH-H_2O]^-$ <br> 329.1392 $[M-H-hexosyl-CH_3COOH-2CH_3]^-$ | Dimethoxy-hexosyl-lariciresinol |
| 116 | 19.82 | – | | | $C_{28}H_{39}O_{13}$ | | 581.2201 | 3.3 | $C_{28}H_{37}O_{13}$ | 359.1445 $[M-H-hexosyl-CH_3COOH]^-$ <br> 329.1392 $[M-H-hexosyl-CH_3COOH-2CH_3]^-$ | Dimethoxy-hexosyl-lariciresinol |
| 117 | 16.37 | – | | | $C_{34}H_{49}O_{18}$ | | 743.2742 | 2.0 | $C_{34}H_{47}O_{18}$ | 581.2249 $[M-H-hexosyl]^-$ <br> 359.1494 $[M-H-2hexosyl-CH_3COOH]^-$ <br> 341.1383 $[M-H-2hexosyl-CH_3COOH-H_2O]^-$ <br> 329.1392 $[M-H-2hexosyl-CH_3COOH-2CH_3]^-$ | Dimethoxy-dihexosyl-lariciresinol |

[a] *Fragment ions produced in MS were named according to Ma et al. [27].*

[b] *Abbreviations:* Caffeic: caffeic acid; Calquin: caffeoylquinic acid; Caftar: caffeoyltartaric acid; DiHBZ: dihydroxybenzoic acid; DiHBZhex: dihydroxybenzoic acid-hexoside; DihydroCaf: dihydrocaffeic acid; Gallic: gallic acid; HBZ: hydroxybenzoic acid; hex: hexose; 4-hydroxyphenylacetic, 4-hydroxyphenylacetic acid; 4-hydroxyphenylacetichex, 4-hydroxyphenylacetic acid-hexoside; Malic: malic acid; pCoumaric: *p*-coumaric acid; Quin: quinic acid; Tartaric: tartaric acid; sh: shoulder.

[c] *Abundances of the fragment ions of caffeoylquinic acids in the negative mode are given in parenthesis.*

unambiguously as *trans*-5-caffeoylquinic acid by comparison with its standard: the deprotonated molecule [M–H]⁻ at $m/z$ 353 yielded fragment ions at $m/z$ 191, 173 and 135; and the protonated molecule [M+H]⁺, at $m/z$ 163 and 145. Moreover, its sodium adducts, [M+Na]⁺ and [2M+Na]⁺ at $m/z$ 377 and 731 respectively, were also observed. Compounds 1 (Rt = 4.74 min, λmax = 301, 323 nm) and 6 (Rt = 10.23 min, λmax = 301, 316 nm) had the same fragmentation pattern as 5-CQA, and their $m/z$ values for [M+H]⁺ and [M–H]⁻ were confirmed with the sodium adduct at $m/z$ 377 in positive ionization mode, and the [2M–H]⁻ ion at $m/z$ 707 in negative mode. All three peaks (1, 3, 6) yielded the same base peak at $m/z$ 191 due to the deprotonated quinic moiety in the negative high-energy function. None of the peaks yielded an intense fragment ion at $m/z$ 173 ([quinic acid-H-H₂O]⁻). This dehydrated ion of quinic acid is characteristically formed in the negative ion mode when the cinnamoyl group is bonded to the quinic moiety at position 4, as already noted by other authors using other QqQ/MS or IT/MS [37, 38]. Peak 1 also gave intense ions from the caffeoyl moiety ([caffeic acid-H-CO₂]⁻) at $m/z$ 135 (71% relative abundance (RA)) and ([caffeic acid-H]⁻) at $m/z$ 179 (32% RA), characteristic intense ions of the fragmentation pattern of 3-CQA by QqQ/MS [37]. The relative hydrophobicity of cinnamoyl derivatives depends on the position, the number, and the identity of the cinnamoyl residues. In general, those CGAs with a greater number of free equatorial hydroxyl groups in the quinic acid are more hydrophilic than those with a greater number of free axial hydroxyl groups [39]. Taking into account the fact that the hydroxyl groups in the quinic acid are axial in position 1 and 3, and equatorial in positions 4 and 5, the elution order observed for monoacyl-CGAs on end-capped C18 reversed-phase LC columns is 3-CGA, 5-CGA, and 4-CGA. This empirical rule was observed by several authors [23, 37, 38, 40]. Therefore, isomers substituted in position 3 were the most hydrophilic; and in position 4, the most hydrophobic, although in some packings 4-CQA precedes 5-CQA. On the other hand, the ease of removal of the caffeoyl residue during fragmentation is 1 ≈ 5 > 3 > 4 [39]. In the negative low energy function, the base peaks were [M–H]⁻ at $m/z$ 353 for peak 1, and [quinic acid-H]⁻ at $m/z$ 191 for peaks 3 and 6, revealing that the caffeoyl moiety in peak 1 was bonded to the quinic structure in a stronger position. So, peak 1 was tentatively assigned to a 3-CQA isomer.

Besides the three major peaks (1, 3, 6), other four caffeoylquinic acid isomers (2, Rt = 6.65 min; 4, Rt = 8.12 min; 5, Rt = 8.36 min; 7, Rt = 15.06

min) were detected in the chromatograms extracted at *m/z* 353 (ESI–) and 355 (ESI+), presenting the same fragmentation pattern in the positive mode as the former isomers. Chlorogenic acid isomers 1-CQA, 3-CQA (neochlorogenic acid), *cis*-3-CQA, 4-CQA (cryptochlorogenic acid), *cis*-4-CQA and *cis*-5-CQA have been previously found in different *Asteraceae* species [41, 42]. In the negative low energy function, compounds **2**, **4** and **7** yielded the deprotonated molecule [M–H]⁻, whereas all four peaks presented the same base peak at *m/z* 191 due to the deprotonated quinic moiety in the negative high energy function. Furthermore, peak **4** yielded ions at *m/z* 135 (21% RA) and at *m/z* 179 (12% RA); and peak **5**, at *m/z* 173 (13% RA), whereas for all other isomers, this ion was less than 4% RA. Peak **5**, presenting the most intense *m/z* 173 and eluting later than 5-CQA (**3**), was ascribed to a 4-CQA isomer.

It is widely accepted that *trans* isomers are the substrates and products of the main phenylpropanoid biosynthetic pathway, being the predominant species detected in plant tissues. However, it is also known that conversion to the *cis* form occurs readily, especially after exposure to UV light, and therefore *cis* isomers might reasonably be expected in plant extracts [43]. Indeed, *cis*-3-CQA, *cis*-4-CQA, and *cis*-5-CQA have been previously found in different *Asteraceae* species [39, 41, 42]. *Cis* isomers fragment identically to the more common *trans* isomers; however, *cis* and *trans* isomers are easily resolved by chromatography. *Cis*-5-acyl and *cis*-1-acyl CGAs are more hydrophobic, thus elute later than their *trans* isomers, whereas the opposite happens with *cis*-3-acyl and *cis*-4-acyl CGAs on end-capped C18 and phenyl hexyl packings [43]. These observations helped to tentatively identify some compounds. Thus, peak **6** was attributed to *cis*-5-CQA, taking into account the elution order of *cis* and *trans* isomers; the fact that absorption maximum for *cis*-CGA occurs at a shorter wavelength than for their *trans* form [44]; and that it is a major peak as its *trans* isomer. Peaks **1** and **4**, which showed similar fragmentation patterns, were designated to the *trans* and *cis* isomers of 3-CQA, respectively.

Peak **2** showed a similar fragmentation pattern to peaks **3** and **6**. Indeed, 1-CQA and 5-CQA are not possible to be reliably distinguished by their fragmentation [39]. Fortunately, *trans*-5-CQA is readily available from commercial sources, and 1-CQA can be easily resolved in the chromatographic elution from this, so, in practice, discrimination is straightforward. Peak **2** eluted earlier than *trans*-5-CQA (**3**) and was assigned to a 1-acyl isomer. The remaining peak (**7**) eluted the latest of all CQA; therefore, it was ascribed to the other 4-CQA isomer.

Taking into account all the above considerations, the chromatographic peaks were tentatively identified as: **1**, *trans*-3-CQA; **2**, *trans*-1-CQA; **3**, *trans*-5-CQA; **4**, *cis*-3-CQA; **5**, *trans*-4-CQA; **6**, *cis*-5-CQA; and **7**, *cis*-4-CQA. Only three CQA isomers had been reported previously in green lettuce, i.e., 5-CQA, 3-CQA and an unidentified CQA isomer and only 5-CQA in the red oak-leaf cultivar [16, 18, 23, 33, 45]. *trans*-5-CQA (**3**) was the major phenolic compound in all the studied lettuce cultivars, as occurs in other lettuce varieties [18, 46, 47]. The following major CQAs were *cis*-5-CQA and *trans*-3-CQA (20% and 8% of the total intensity of *trans*-5-CQA respectively in the butterhead cultivar, 40% and 5% in the green oak-leaf cultivar, and about 5% for each CQA isomer in the red oak-leaf cultivar).

### 7.3.1.1.2 *p-Coumaroylquinic Acids*

Compounds **8** (Rt = 9.82 min, max = 312 nm) and **9** (Rt = 13.74 min, max = 308 nm) were identified as *p*-coumaroylquinic acid isomers on the basis of mass spectral data and UV spectra, which followed the pattern of the *p*-coumaric acid standard. In both low and high energy positive ion mode, the sodium adduct [M+Na]$^+$ at $m/z$ 361 was the base peak for both compounds, and the ion at $m/z$ 147 ([p-coumaroyl+H]$^+$) was the secondary most intense ion. In the negative low energy function, the base peaks were [M–H]$^-$ at $m/z$ 337 for peak **8**, and [quinic acid-H]$^-$ at $m/z$ 191 for peak **9**, revealing that the *p*-coumaroyl moiety in peak **8** was bonded to the quinic structure in a stronger position. Moreover, peak **8** yielded in the high energy function an intense ion at $m/z$ 119 due to its decarboxylation product [p-coumaric acid-H-CO$_2$]$^-$, which is characteristic of the fragmentation pattern of 3-*p*-coumaroylquinic acid, thus this isomer was tentatively assigned to peak **8**, for the first time in lettuce cultivars. The base peak of compound **9** at $m/z$ 191 due to the deprotonated quinic moiety is characteristic of 5-*p*-coumaroylquinic acid [38]. Similarly to CQA isomers, the elution order of both isomers on end-capped C18 packings agrees with these tentatively assignments. 5-*p*-coumaroylquinic acid and an unidentified isomer have been previously reported in bibliography in green lettuce cultivars [23, 46]. To the authors' knowledge the presence of *p*-coumaroylquinic acids are here reported in green and red oak-leaf cultivars for the first time.

### 7.3.1.1.3 Caffeoyltartaric Acid

A caffeoyltartaric acid (peak **10**: Rt = 9.06 min, max = 301, 323 nm) was detected in the extracted MS chromatogram set at $m/z$ 311 in the negative ion mode, presenting the corresponding fragmentation pattern: The dehydrated protonated molecule at $m/z$ 293 was the base peak in low energy function; and intense fragments of the deprotonated tartaric ($m/z$ 149) and caffeic ($m/z$ 179) acids and the losses of water ($m/z$ 293) and $CO_2$ ($m/z$ 135; base peak) were observed in the high energy function. Two isomers of caffeoyltartaric acid have been already reported in lettuce in literature [23, 34, 45, 46, 48].

### 7.3.1.1.4 p-Coumaroyltartaric Acid

Peak **11** (Rt = 15.63 min, max = 310 nm), detected in the extracted MS chromatogram set at $m/z$ 295 in the negative ion mode, yielded the base peak at $m/z$ 163 due to the deprotonated *p*-coumaric acid, and two fragments at $m/z$ 149 (50% RA) and $m/z$ 119 (60% RA) due to the deprotonated tartaric acid and the descarboxilation of *p*-coumaric acid in the low energy function. Thus, compound **11** was tentatively identified as *p*-coumaroyltartaric acid, which has been previously found in green lettuce cultivars being detected here in oak-leaf cultivars for the first time [23, 46].

### 7.3.1.1.5 Caffeoylmalic Acid (CMA)

Caffeoylmalic acid (CMA) (peak **12**: Rt = 9.05 min, max = 301, 323 nm) was detected when the $m/z$ value for the extracted MS chromatogram was set at $m/z$ 295 (negative ion mode) or 297 (positive ion mode). Besides the UV spectra of peak **12** followed the pattern of caffeic acid standard. In the negative ion mode, the high-energy function provided ions corresponding to malic acid: the base peak at $m/z$ 133 was due to the deprotonated malic moiety; and fragment ions, to the losses of water and CO at $m/z$ 115 and 105 respectively. $MS^E$ experiments in the positive ion mode showed that CMA behaved as described above for CQA, yielding the same ions from the caffeoyl moiety, as well as the sodium adduct. CMA, commonly named phaseolic acid, has been described before in different lettuce cultivars but detected in green oak-leaf lettuce for the first time in the present work [23, 34, 46, 48].

### 7.3.1.1.6 Dicaffeoylquinic Acids and Caffeoylquinic Acid Glycosides

Both dicaffeoylquinic acids (diCQA) and caffeoylquinic acid-hexosides present an average molecular mass of 516 $u$, and produce isobaric deprotonated or protonated molecules at $m/z$ 515 and 517 in the negative and positive ion modes, respectively. Five peaks were detected in the extracted MS chromatograms at these $m/z$ values: peak **13** (Rt = 5.86), peak **14** (Rt = 7.56), peak **15** (Rt = 20.20, max = 321 nm), peak **16** (Rt = 20.63, max = 326 nm) and peak **17** (Rt = 24.17, max = 331 nm). Based on their accurate masses and fragmentation patterns, these peaks were distinguished as either di-caffeoylquinic acids (**15**, **16** and **17**) with monoisotopic [M–H]⁻ at $m/z$ 515.1190 ($C_{25}H_{23}O_{12}$) and monoisotopic [M+H]⁺ at $m/z$ 517.1346 ($C_{25}H_{25}O_{12}$), and caffeoylquinic acid-hexosides (**13** and **14**) with monoisotopic [M–H]⁻ at $m/z$ 515.1401 ($C_{22}H_{27}O_{14}$) and monoisotopic [M+H]⁺ at $m/z$ 517.1548 ($C_{22}H_{29}O_{14}$), in the negative and positive ion modes, respectively.

The first fragments of the diCQA were due to the loss of one of the caffeoyl moieties, leading to the precursor ion of a CQA; therefore, subsequent fragmentation of these ions yielded the same fragments as the corresponding CQA. In the positive low energy function, the sodium adducts at $m/z$ 539 and the dehydrated protonated molecule at $m/z$ 499 were detected with different % RA: peak **15**, [M+H-H₂O]⁺ base peak and [M+Na]⁺ 80% RA; peak **16**, [M+Na]⁺ base peak and [M+H-H₂O]⁺ 20% RA; and peak **17**, [M+Na]⁺ base peak and [M+H-H₂O]⁺ 90% RA. The positive high energy function gave a base peak at $m/z$ 163 ([caffeic acid+H-H₂O]⁺) for the three peaks, but [M+Na]⁺ presented 50% RA for peak **15**, 35% RA for peak **16**, and 70% RA for peak **17**. The % RA differences between these ions are related to the difficulty of removing the acylating residue at different positions. In accordance with this, the negative low energy function MS spectra disclosed that peak **17** yielded only the deprotonated molecule ($m/z$ 515) as the base peak; peak **15**, the base peak [M–H]⁻ and the fragment [CQA-H]⁻ ion at $m/z$ 353 with 65% RA; and peak **16**, the base peak [CQA-H]⁻ at $m/z$ 353 and [M–H]⁻ with 40% RA. Hence, these observations suggest that peak **17** contains a caffeoyl moiety at the positions more difficult to be removed (4 > 3 > 5 ≈ 1) [38, 39] than the other peaks, followed by peak **15**. Indeed, the presence of the dehydrated quinic residue ion [quinic acid-H-H₂O]⁻ at $m/z$ 173 as the base peak in the high negative energy spectra of peak **17** revealed that one of the caffeoyl moieties was bonded to quinic acid at position 4. Then it remained to be determined if the other caffeoyl moiety was substituted at position 1, 3 and 5. Finally, taking also into account the elution order of diCQA isomers (RT on end-capped C18 packings: 1,3-diCQA <<< 1,4-diCQA << 3,4-diCQA < 1,5-diCQA < 3,5-diCQA << 4,5-diCQA) reported in bibliography, compound **17** was assigned to 4,5-diCQA. In the high negative energy function, base peaks of compounds **15** and **16** were [quinic acid-H]⁻ at *m/z* 191, whereas the characteristic fragment at *m/z* 173 corresponding to the dehydrated quinic residue ion was not detected [37, 39]. Therefore, caffeoyl moieties were substituted at position 1, 3 and 5. Compound **15** was identified unambiguously as 1,5-diCQA by comparison with its standard. Thus, regarding its RT and the ease of removal of the caffeoyl residue, compound **16** was assigned to 3,5-diCQA. Isomers 3,5-diCQA (isochlorogenic acid A), *cis*-3,5-diCQA, and 4,5-diCQA (isochlorogenic acid B) have previously been reported in *L. sativa* [18, 23, 46, 48]. Among these, isochlorogenic acid A was reported to be the most abundant in lettuce, as found in the present study, which supported the assignment of compound **16** [20, 45, 49]. 1-acyl CGA has been found in some *Asteraceae*; however, the isomer 1,5-diCQA is reported in lettuce here for the first time [39].

Caffeoylquinic acid-hexosides (**13** and **14**) base peaks were their sodium adducts in the positive ion mode and the deprotonated molecule in the negative ion mode, which confirmed their identities. The presence of the fragment ion at *m/z* 353 due to the deprotonated CQA, and the base peak at *m/z* 191 due to the deprotonated quinic acid in the negative high-energy function of peak **13** also support the assignment. Peak **14** was at trace levels, not being possible to register its fragmentation pattern. To the authors' knowledge, caffeoylquinic acid-hexosides have not been reported in lettuce before.

### 7.3.1.1.7 *p-Coumaroylcaffeoylquinic Acids*

Two chromatographic peaks showed protonated and deprotonated molecules that corresponded to *p*-coumaroylcaffeoylquinic acids, at *m/z* 501 in the positive ion mode and at *m/z* 499 in the negative mode: peak **18** (Rt = 23.58 min, max = 312 nm) and peak **19** (Rt = 23.95 min, max = 316 nm). In the positive high energy function, the base peaks yielded by both isomers were the fragment ion at *m/z* 147 due to [*p*-coumaroyl+H]⁺, disclosing that the *p*-coumaroyl moiety was attached to the quinic acid in a weaker position than the caffeoyl one. This was also supported by the fragmentation pattern observed for both peaks in the negative ion mode, which yielded the deprotonated Molecules and fragments at $m/z$ 353 due to the loss of the *p*-coumaroyl moiety (85–95% RA) and at $m/z$ 337 due to the loss of the caffeoyl moiety (40–50% RA) in the low energy function, indicating that the former loss was favored. This fragmentation pattern was reported for 3-*p*-coumaroyl-4-caffeoylquinic acid (3-*p*Co-4-CQA) and 4-caffeoyl-5-*p*-coumaroylquinic acid (4-C-5-*p*CoQA) [50]. The deprotonated quinic acid ion at $m/z$ 191 was the base peak in the high-energy function; this fragment is a characteristic base peak of 5-CQA, 3-CQA and 5-*p*CoQA, and is yielded by 4-CQA [38]. Thus, taking also into account that the elution order on end-capped C18 packings is 3,4-isomers, 3,5-isomers and 4,5-isomers compounds **18** and **19** were tentatively assigned to 3-*p*Co-4-CQA and 4-C-5-*p*CoQA, respectively, for the first time in lettuce cultivars. *p*-Coumaroylcaffeoylquinic acids have been previously reported in lettuce but is the first time in green and red oak-leaf cultivars [23, 42, 50].

### 7.3.1.1.8   *Dicaffeoyltartaric Acids*

Two peaks (20, 21), presenting the same UV spectra as caffeic acid standard, were detected in the chromatograms extracted from the TIC MS scan chromatogram in positive and negative modes at $m/z$ 475 and 473, respectively, which were due to two dicaffeoyltartaric acid isomers (diCTA). Compound 20 (Rt = 10.53 min, λmax = 301, 324 nm) and compound 21 (Rt = 12.54 min, λmax = 301, 323 nm) presented the same fragmentation pattern, and their identity was confirmed with the sodium adduct at $m/z$ 497 in positive ionization mode and the [2M–H]⁻ ion at $m/z$ 947 in negative mode for peak 20, and the protonated and deprotonated molecules for peak 21. In the negative ion mode, both peaks (20, 21) yielded the same base peak at $m/z$ 293 due to the loss of water of the deprotonated caffeoyltartaric acid, and [CTA-H]⁻ at $m/z$ 311 due to the loss of one of the caffeoyl moieties, as well as ions from the tartaric moiety, [tartaric acid-H]⁻ at $m/z$ 149 and [tartaric acid-H-CO$_2$]⁻ at $m/z$ 105; and ions from the caffeoyl moiety, [caffeic acid-H]⁻ at $m/z$ 179 and [caffeic acid-H-CO$_2$]⁻ at $m/z$ 135. Compound 20 was tentatively identified as di-*O*-caffeoyltartaric (chicoric acid), and compound 21 as meso-di-*O*-caffeoyltartaric acid, since they were detected in lettuce elsewhere; the former being reported as the most abundant as we observed [16–18, 20, 23, 32, 34, 45, 46, 48–50].

### 7.3.1.1.9 Other Hydroxycinnamic Acid Derivatives

Several cinnamoyl glycosides were found in the lettuce extracts, such as caffeoyl-hexosides, *p*-coumaroyl-hexosides, sinapoyl-hexosides, and dihydrocaffeic acid-hexosides, whose fragmentation patterns were characterized by the aglycone product ion resulted from the loss of a hexose residue [23, 52].

Eight peaks (22, Rt = 5.39 min; 23, Rt = 5.64 min; 24, Rt = 6.08 min, $\lambda$max = 301, 325 nm; 25, Rt = 7.69 min; 26, Rt = 8.44 min; 27, Rt = 9.01 min; 28 Rt = 9.52 min; and 29 Rt = 9.64 min) were observed in the chromatogram extracted at *m/z* 343 and 341 in positive and negative ion modes respectively. All of them (22–29) produced *m/z* 179 and 135 in negative ion mode, and *m/z* 163, 145, 135, 117, and 89 in positive ion mode, consistent with the presence of a caffeic acid residue. Thus, these compounds were tentatively assigned as isomeric caffeic acid-hexosides, in agreement with Clifford et al. [41]. Moreover, the identity of peaks 22–26 and 28 were confirmed by the presence of their sodium adducts in the positive low energy function. As well, peak 30 (Rt = 8.01 min, $\lambda$max = 301, 325 nm) showed the same fragmentation pattern as caffeic acid, yielding also a monoisotopic protonated molecule at *m/z* 359.0802 ($C_{18}H_{15}O_8$) in the positive ion mode, and a monoisotopic deprotonated molecule at *m/z* 357.0633 ($C_{18}H_{13}O_8$) in the negative ion mode. Thus, it was tentatively assigned as a caffeoyl derivative; however, the nature of the non-phenolic residue (196.0387 *u*) was not able to be disclosed. Such caffeoyl derivative has not previously been reported in lettuce so far we are aware.

Similarly, four isomers of synapic acid-hexosides (**31**, Rt = 6.03 min, $\lambda$max = 301, 326 nm; **32**, Rt = 9.70 min; **33**, Rt = 10.36 min; **34**, Rt = 13.13 min) were tentatively identified in the extracted traces at *m/z* 387 and 385 in the positive and the negative ion modes respectively. Ions corresponding to the deprotonated aglycone at *m/z* 223, and the subsequent decarboxylations and losses of methyl residues at *m/z* 208, 179, 164, and 149 from the synapoyl moiety were detected in the negative ion mode. In addition, the positive ion mode yielded the sodium adduct at *m/z* 409 and ions due to the loss of the hexose residue at $m/z$ 225, and subsequent losses of $H_2O$ at $m/z$207, $CH_3OH$ at $m/z$ 192, and CO at $m/z$ 129. One isomer of synapic acid-hexoside has been previously reported in green lettuce cultivars [23].

Following this fragmentation patterns, a *p*-coumaric acid-hexoside (35, Rt = 8.32 min) and two dihydrocaffeic acid-hexosides (36, Rt = 3.70 min; 37, Rt = 3.83 min) were also characterized. All of them yielded the product ion due to the loss of the hexose residue ($m/z$ 163 for 35, $m/z$ 181 for 36 and 37), with the subsequent losses of $H_2O$, CO, and $CO_2$ in the negative ion mode; and the sodium adduct in the positive ion mode ($m/z$ 349 for 35, $m/z$ 367 for 36 and 37).

Seven caffeic acid-hexosides, a synapic acid-hexosides, a dihydrocaffeic acid-hexoside and a *p*-coumaric acid-hexoside have been previously reported in green lettuce cultivars [23]. In the present work, one more caffeic acid-hexoside, a dihydrocaffeic acid-hexoside and three synapic acid-hexosides were identified in the three lettuce cultivars studied.

Peaks 38 (Rt = 11.81 min, λmax = 307 nm), 39 (Rt = 14.47 min) and 40 (Rt = 16.48 min) were tentatively proposed as isomers of ferulic acid methyl esters. According to previous data [23, 36], these compounds showed demethylated fragment ions at $m/z$ 192 ([M-H-$CH_3$]$^-$) and $m/z$ 177 ([M-H-2$CH_3$]$^-$), which is characteristic of the methoxylated cinnamic acids. Two of these isomers of ferulic acid methyl esters have been previously reported in green lettuce cultivars [23].

## 7.3.1.2   HYDROXYBENZOIC DERIVATIVES

Hydroxybenzoic derivatives were not detected in the positive ion mode. Thus, no peaks were detected in the chromatograms extracted from the TIC MS scan chromatogram at the protonated molecule or the sodium adduct masses of the hydroxybenzoic derivatives observed in the negative ion mode. Only one of the two previously reported in green lettuce cultivars [23] isomers of hydroxybenzoic acid (**41**: Rt = 4.67 min) and dihydroxybenzoic acid (**42**: Rt = 5.42 min) were detected at $m/z$ 137 and $m/z$ 153, respectively. Their corresponding decarboxylated ions were also observed at $m/z$ 93 and $m/z$ 109, respectively. The dihydrobenzoic acid (**42**) was only detected in the butterhead and green oak-leaf lettuce cultivars.

Several hydroxybenzoic glycoside esters were characterized according to their MS data and fragmentation pattern by the neutral loss of the glycosidic moiety. Hydroxybenzoic acid-hexosides (**43**, Rt = 4.22 min; **44**, Rt = 5.15 min) yielded the deprotonated ion at $m/z$ 299 and the product ions due to losses of the hexose residue ($m/z$ 137) and $CO_2$ ($m/z$ 93). Dihydroxybenzoic acid-hexosides (**45**, Rt = 2.49 min; **46**, Rt = 2.69 min; **47**, Rt = 3.74 min; **48**, Rt = 3.91 min; **49**, Rt = 4.48 min; **50**, Rt = 4.68 min) produced the deprotonated molecule at $m/z$ 315 (base peak), an odd electron product ion at $m/z$ 152 corresponding to the loss of hexose plus H (163 $u$), an even electron ion at $m/z$ 153 due to the loss of hexose, the dehydrated ion at $m/z$ 135, and the decarboxylated ion at $m/z$ 109, in agreement with bibliography [23]. Hence, one more hydroxybenzoic acid-hexoside and four more dihydroxybenzoic acid-hexosides are here detected in lettuce than in previous studies on different lettuce cultivars. The release of such unusual losses was also observed for gallic acid-hexoside isomers. Thus, peaks **51** (Rt = 2.80 min), **52** (Rt = 2.88 min) and **53** (Rt = 6.61 min) were tentatively proposed as gallic acid-hexosides, since they yielded the deprotonated molecule at $m/z$ 331 (base peak), and an odd electron product ion at $m/z$ 168, corresponding to the loss of hexose plus H (163 $u$), an even electron ion at $m/z$ 169 due to the loss of hexose, and [gallic acid-H-$CO_2$]$^-$ at $m/z$ 125. Two isomers of gallic acid-hexoside have been detected previously only in the lettuce cv. baby [23].

Aside from the loss of the hexose moiety, syringic acid-hexoside (**54**, Rt = 5.90 min, $m/z$ 359) showed subsequent losses of $CH_3$ from the methoxy groups of the aglycone and $CO_2$ ($m/z$ 182, 153, 138 and 123), as previously observed in literature [23, 52].

In agreement with previous studies [23], compounds **55** (Rt = 17.09 min) and **56** (Rt = 24.83 min) showing a deprotonated molecule at $m/z$ 451 were tentatively assigned as hydroxybenzoyl-gallic acid-hexosides. The high-energy function yielded the fragment ion corresponding to the deprotonated gallic acid-hexoside at $m/z$ 331, after the loss of the hydroxybenzoyl moiety (120 $u$). As well, product ions due to successive losses of $H_2O$ at $m/z$ 313, hexose plus H at $m/z$ 168 and $CO_2$ at $m/z$ 124 were observed. A similar pattern was found for the hydroxybenzoyl-dihydroxybenzoic acid-hexosides (**57**, Rt = 17.68 min; **58**, Rt = 19.41 min; **59**, Rt = 23.64 min; **60**, Rt = 26.88 min, λmax = 256, 335 nm; **61**, Rt = 27.09 min) detected in the extracted trace at $m/z$ 435. For peak **59**, only the deprotonated molecule was detected due to its low concentration in the extract. All other isomers yielded the fragment ions corresponding to [dihydroxybenzoic acid-hexoside-H]$^-$ at $m/z$ 315, and the subsequent losses of $H_2O$ at $m/z$ 297 and hexose plus H at $m/z$ 152 and $CO_2$ at $m/z$ 108. Peaks **58** and **61** showed the product ion [dihydroxybenzoic acid-H]$^-$ due to an even electron ion at $m/z$ 153 (loss of hexose), instead of the odd electron product ion at $m/z$ 152. Besides, peaks **57**, **60** and **61**, yielded the fragment ion [hydroxybenzoic acid-H]$^-$ at $m/z$ 137 and its corresponding decarboxylation ion at $m/z$ 93. This behavior agrees with that observed for hydroxycinnamic acid glycosides above and in literature [41], which suggest that both, the hydroxybenzoic acid moiety and the dihydroxybenzoic acid moiety, are attached through their phenolic hydroxyl to different positions of the same hexose molecule. Just one isomer of hydroxybenzoyl-gallic acid-hexoside and two isomers of hydroxybenzoyl-dihydroxybenzoic acid-hexosides have been previously characterized only in cv. baby lettuce [23].

### 7.3.1.3   HYDROXYPHENYLACETIC DERIVATIVES

Taking into account the MS data, the fragmentation patterns observed for hydroxybenzoic acid in the negative ion mode and bibliography [23, 51], 4-hydroxyphenylacetic acid was tentatively assigned to peak **62** (Rt = 5.60 min), which yielded the deprotonated molecule at $m/z$ 151 and fragment ions due to the loss of CO at $m/z$ 123 and $CO_2$ at $m/z$ 107, showing the typical decarboxylation of PA. Likewise, peak **63** (Rt = 5.20 min, λmax = 270, 276 nm) observed in the extracted trace at $m/z$ 313, produced the same decarboxylation ions, and a fragment ion at $m/z$ 151 due to deprotonated 4-hydroxyphenylacetic acid obtained after the loss of a hexose moiety. Thus, it was proposed as 4-hydroxyphenylacetic acid-hexoside. Both compounds have been previously detected in green lettuce cultivars [23].

### 7.3.2   FLAVONOIDS

### 7.3.2.1   FLAVONOLS

Thirteen quercetin glycosides (**64–76**) and four kaempferol glycosides (**77–80**) were detected and identified on the basis of their mass spectral data, comparison with available standards, and literature. Flavonol monoglycoside mass spectra in the positive mode showed the protonated molecule $[M+H]^+$, the sodium adduct ion $[M+Na]^+$ and the protonated aglycone ion $[Y_0]^+$ as a result of the loss of the sugar or organic acid residue (losses:146

*u*, rhamnosyl residue; 162 *u*, hexosyl residue; 176 *u*, glucuronic residue; 178 *u*, gluconic residue; 248 *u*, malonyl-hexosyl residue; 324 *u*, di-hexosyl residue; 338 *u*, glucuronic + hexosyl residue; 410 *u*, hexosyl + malonyl-hexosyl residue; 424 *u*, glucuronic + malonyl-hexosyl residue). In the mass spectrum of flavonol diglycosides, a fragment $[Y_1]^+$ due to the loss of the first sugar or organic acid unit was also observed. In the negative mode, the high-energy function product ions corresponding to quercetin at *m/z* 300 (odd electron ion) and/or 301 (even electron ion) were detected, as observed in MS/MS elsewhere [23]. Regarding this, compounds **64** (Rt = 17.16 min, λmax = 279, 344 nm), **65** (Rt = 18.03 min, λmax = 252, 367 nm) and **66** (Rt = 20.25 min, λmax = 252, 330 nm) were identified as quercetin-3-*O*-hexosides on the basis of their protonated molecule at *m/z* 465 and a high energy function product ion at *m/z* 303, which indicates cleavage of a hexosyl group. This fragmentation pattern and chromatographic RT of the reference standard confirmed that compound **66** was quercetin-3-*O*-galactoside. Two isomers of quercetin hexose have been previously described in lettuce [16–18, 20, 23, 31, 33, 34, 45, 48, 49].

Compound **67** (Rt = 18.44 min, λmax = 254, 349 nm) was identified as quercetin-3-*O*-glucuronide because of $[M+H]^+$ at *m/z* 479, $[M+Na]^+$ at *m/z* 501 and $[Y_0]^+$ at *m/z* 303, which indicated the loss of a glucuronic residue in the positive mode. Similarly, in the negative mode, the deprotonated molecule $[M–H]^-$ at *m/z* 477 yielded $[Y_0]^-$ at *m/z* 301; the loss of 176 *u* pointed out the presence of a glucuronic residue. This glucuronic group was also observed in compound **68** (Rt = 9.50 min, λmax = 256, 352 nm) and compound **69** (Rt = 10.58 min), which gave $[M+H]^+$ at *m/z* 641, $[M+Na]^+$ at *m/z* 663, and $[Y_0]^+$ at *m/z* 303 in positive mode, and peak **69**, also $[Y_1]^+$ at *m/z* 465. In the negative mode, both compounds presented similar ionization and fragmentation pattern: $[M–H]^-$ at *m/z* 639, $[Y_1]^-$ at *m/z* 463 and $[Y_0]^-$ at *m/z* 300 (odd electron ion) and/or 301 (even electron ion). Moreover, the loss of 162 *u* revealed the cleavage of a hexoxyl group, therefore, these flavonols were assigned to quercetin hexose-glucuronide isomers, which had been already described in baby, romaine, and iceberg cultivars, but are reported here for the first time in the green and red oak-leaf cultivars.

Compounds **70** (Rt = 21.52 min, λmax = 255, 352 nm), **71** (Rt = 22.03 min, λmax = 252, 364 nm) and **72** (Rt = 23.69 min) were identified as quercetin malonylhexoside isomers since they presented $[M+H]^+$ at *m/z* 551, $[M+Na]^+$ at *m/z* 573, and $[Y_0]^+$ at *m/z* 303 due to the loss of the malonylhexosyl moiety in the positive ion mode; and $[M–H]^-$ at $m/z$ 549, $[Y_0]^-$ at $m/z$ 301, $[M–H–CO_2]^-$ at $m/z$ 505 (base peak) in the negative ion mode. The neutral loss of $CO_2$ is characteristic of compounds presenting the malonyl group, as previously reported [23]. This fact is due to in-source fragmentation, which can affect the correct identification of the deprotonated molecule of interest, because the RA of $[M–H]^-$ ion could be lower than the product ion $[M–H–CO_2]^-$ as occurred with these peaks. This particularly labile group could be partially lost during ion transfer from a higher-pressure region of the source to a lower-pressure region [53], as observed for peak **70** (0.4% RA), peak **71** (11% RA) and peak **72** (0.4% RA). The identification of compound **70** was also confirmed by the presence of $[2M–H]^-$ ion. Quercetin-3-$O$-(6"-$O$-malonyl)-glucoside has been reported in lettuce in several publications [16, 18, 20, 33, 34, 46, 49, 51]. Two isomers of quercetin malonylglucoside were already described in different lettuce varieties [23, 48]. The presence of three quercetin malonylhexoside isomers in lettuce is described for the first time in the present study.

Compound **73** (Rt = 11.51 min, λmax = 253, 355 nm) was identified as quercetin-3-$O$-(6"-$O$-malonyl)-glucoside-7-$O$-glucuronide, which has been previously described in lettuce [18, 23, 34], however, it is reported here in green oak-leaf lettuce for the first time as far as we know. In the positive ion mode, $[M+H]^+$ at $m/z$ 727, $[M+Na]^+$ at $m/z$ 749, and the fragment ions $[Y_1]^+$ at $m/z$ 479 and $[Y_0]^+$ at $m/z$ 303 indicated the loss of a malonyl-glucosyl group followed by a glucuronic group. In the negative ion mode, the neutral loss of $CO_2$ yielding $[M–H–CO_2]^-$ at $m/z$ 681 confirmed the presence of a malonyl residue in the molecular structure; as well as the high energy function product ions at $m/z$ 300 (odd electron ion) and/or 301 (even electron ion), the presence of quercetin. Similarly, compound **74** (Rt = 13.82 min, λmax = 253, 350 nm) also contained a malonyl residue since its base peak in the negative mode was $[M–H–CO_2]^-$ at $m/z$ 667. The deprotonated molecule at $m/z$ 711 was also present and $[Y_0]^-$ at $m/z$ 300 (odd electron ion) and/or 301 (even electron ion) indicated that the aglycone was quercetin. The positive ion mode yielding $[M+H]^+$ at $m/z$ 713, $[M+Na]^+$ at $m/z$ 735, and the fragment ions $[Y_1]^+$ at $m/z$ 465 and $[Y_0]^+$ at $m/z$ 303 confirmed the cleavage of malonylhexosyl group followed by a hexosyl group. Thus, compound **74** was tentatively assigned to quercetin-3-$O$-(6"-$O$-malonyl)-glucoside-7-$O$-glucoside, which has been previously reported in lettuce [23, 34].

Compounds **75** (Rt = 12.18 min) and **76** (Rt = 16.07 min) presented the same monoisotopic molecular mass for $[M+H]^+$ at $m/z$ 627.1580 $(C_{27}H_{31}O_{17})$ and $[M-H]^-$ at $m/z$ 625.1405 $(C_{27}H_{29}O_{17})$, and $[M+Na]^+$ at $m/z$ 649.1381 $(C_{27}H_{30}O_{17}Na)$. The presence of $[Y_0]^+$ at $m/z$ 303 and $[Y_0]^-$ at $m/z$ 301 in the positive and negative ion modes, respectively, disclosed that the aglycone was quercetin. However, these compounds followed different fragmentation patterns. Peak **75** yielded $[Y_1]^-$ at $m/z$ 463 due to the loss of a hexosyl moiety (162 $u$), and revealing that $[Y_0]^-$ was obtained from the loss of a second hexosyl residue. Thus, compound **75** was assigned as a quercertin-$O$-di-hexoside. Instead, peak **76** yielded $[Y_1]^-$ at $m/z$ 447 due to the loss of a gluconic moiety (178 $u$), and disclosing a subsequent loss of a rhamnosyl moiety (146 $u$) to achieve $[Y_0]^-$. Peak **75** was tentatively identified as quercetin-di-glucoside, which has been previously reported in green lettuce and ruby red lettuce [34]. Peak **76** was tentatively proposed as quercertin-$O$-rhamnosyl-gluconide, which is here reported for the first time to the author's knowledge.

Regarding kaempferol conjugates, compound **77** (Rt = 25.27 min, λmax = 265, 347 nm) was identified as kaempferol-3-$O$-(6"-$O$-malonyl)-glucoside, which has been already found in different lettuce cultivars [51]. In the positive mode, $[M+H]^+$ at $m/z$ 535, $[M+Na]^+$ at $m/z$ 557, and the fragment ions and $[Y_0]^+$ at $m/z$ 287 revealed the cleavage of a malonyl-glucosyl group. In the negative mode, $[M-H]^-$ at $m/z$ 533, $[Y_0]^-$ at $m/z$ 285, $[M-H-CO_2]^-$ at $m/z$ 489 confirmed the presence of the malonyl glucosyl moiety in the molecule. Regarding the aglycone, kaempferol, and the flavone luteolin are isobaric, but their conjugates can be distinguished on the basis of their $MS^E$ data. In the positive low energy function, kaempferol derivatives yield $[Y_0]^+$ as the base peak or $[M+H]^+$ as the base peak plus an intense $[Y_0]^+$, whereas luteolin derivatives give as the base peak $[M+H]^+$ or $[M+H-H_2O]^+$, and $[Y_0]^+$ does not appear or present low RA. In the negative low energy function, both compounds yield $[M-H]^-$ or $[M-H-CO_2]^-$ (in the case of malonylglycosides) as the base peak, but in the negative high energy function, kaempferol conjugates give the base peak $[Y_0]^-$, whereas luteolin compounds yield the base peak $[M-H]^-$ or $[M-H-CO_2]^-$ and an intense $[Y_0]^-$, or $[Y_0]^-$ as the base peak and an intense $[M-H]^-$ with RA higher than 50% RA. Moreover, several minor monoisotopic product ions at $m/z$217.0501 $(C_{12}H_9O_4)$, 199.0395 $(C_{12}H_7O_3)$, 175.0395 $(C_{10}H_7O_3)$ and 133.0290 $(C_8H_5O_2)$ are characteristic of luteolin, and helps to distinguish it from its kaempferol isomers [23, 52]. In this sense, these fragment ions did not appear in the negative high-energy MS spectra of peak **77**, suggesting that it is a kaempferol derivative. Moreover, this identification was also supported by the base peaks yielded in the positive low energy and the negative high energy functions, $[Y_0]^+$ and $[Y_0]^-$ respectively, as well as its UV-visible spectra, and elution order since kaempferol isomers elute later than luteolin isomers one capped C18 packings.

Two isomers (**78**: Rt = 23.90 min; **79**: Rt = 26.43 min) were detected in the extracted MS chromatogram at $m/z$ 449 and 447 in the positive and negative ion modes, respectively, which yielded the protonated ion, $[M+Na]^+$ at $m/z$ 471 and $[Y_0]^+$ at $m/z$ 287 in the positive ion mode, and the deprotonated molecule and $[Y_0]^-$ at $m/z$ 285 in the negative ion mode; revealing the loss of a hexosyl residue and the presence of kaempferol or luteolin aglycone. The base peaks yielded in the positive low energy and the negative high energy functions were $[Y_0]^+$ and $[Y_0]^-$ respectively, and no characteristic minor product ions of luteolin were detected in the negative high energy function, therefore, the aglycone was tentatively identified as kaempferol. Compound **78** was identified unambiguously as kaempferol-3-*O*-glucoside by comparison with its standard, whereas compound **79** as kaempferol-hexoside. Kaempferol-3-*O*-glucoside is the only kaempferol-hexoside that has been previously detected in several lettuce cultivars [15].

Compound **80** (Rt = 22.34 min, λmax = 265, 332 nm) was identified as kaempferol-3-*O*-glucuronide, which has been previously found in lettuce in literature [45]. This compound yielded $[M+H]^+$ at $m/z$ 463, $[M+Na]^+$ at $m/z$ 485 and $[Y_0]^+$ at $m/z$ 287 in the positive mode; and $[M–H]^-$ at $m/z$ 461 and $[Y_0]^-$ at $m/z$ 285 in the negative mode. The observed loss of 176 $u$ pointed out the presence of a glucuronic residue. Besides, the presence of the base peaks $[Y_0]^+$ and $[Y_0]^-$ in the positive low energy and the negative high energy functions, respectively, and the absence of luteolin characteristic minor product ions in the negative high energy function, supports the proposed identification for this compound.

Peak **81** (Rt = 27.08 min) presented the protonated and deprotonated molecules at $m/z$ 287 and 285 in the positive and the negative ion modes, respectively, which yielded fragment ions characteristics of kaempferol or luteolin aglycones [7], suggesting that both compounds were eluting overlapped in this peak. To the author's knowledge, kaempferol aglycone has not been previously found in lettuce, but in escarole (*Asteraceae*) [18].

Kaempferol-hexosides (**78** and **79**), kaempferol-3-*O*-glucuronide (**80**) and kaempferol aglycone (**81**) are reported in the green and red oak-leaf cultivars here for the first time as far as we are aware.

### 7.3.2.2 FLAVONES

Five luteolin glycosides (**82–86**) and five apigenin conjugates (**87–91**) were detected and identified on the basis of mass spectral data, comparing with available standards and bibliographic sources. Compound **82** (Rt = 19.82 min, λmax = 255, 347 nm) was identified unambiguously as luteolin-7-*O*-glucoside by comparison with its standard, which showed the deprotonated molecule at $m/z$ 447, [2M–H]⁻ at $m/z$ 895, $[Y_0]^-$ at $m/z$ 285, and luteolin characteristic minor product ions at $m/z$ 217, 199 and 175 in the negative ion mode; and the protonated molecule at $m/z$ 449, [M+Na]⁺ at $m/z$ 471, $[Y_0]^+$ at $m/z$ 287, and intense fragment ions at 153 and 135 in the positive mode. Luteolin-7-*O*-glucoside has been previously described in lettuce cultivars [15, 23, 48].

Compound **83** (Rt = 17.45 min, λmax = 253, 348 nm) was assigned to luteolin-7-*O*-glucuronide regarding the protonated molecule yielded at $m/z$ 463, [M+Na]⁺ at $m/z$ 485 and $[Y_0]^+$ at $m/z$ 287, which revealed the cleavage of a glucuronic residue. In the negative high energy function, compound **83** yielded the corresponding deprotonated molecule at $m/z$ 461, $[Y_0]^-$ at $m/z$ 285, as well as some minor fragment ions at $m/z$ 217, 199, 175, 151 and 133, which distinguished luteolin conjugates from its kaempferol isomers [23, 34]. This identification was supported by its UV-visible spectrum, which followed the luteolin pattern; and its elution order on end-capped C18 packings, glucuronide conjugates elute earlier than their corresponding glucoside ones. Luteolin-7-*O*-glucuronide has been previously reported in lettuce [23, 34, 48].

Compounds **84** (Rt = 20.27 min) and **85** (Rt = 21.17 min, λmax = 268, 351 nm) showed base peaks at $m/z$ 595 ([M+H]⁺) in the low energy function. Aside, compound **85** also presented the sodium adduct ($m/z$ 617), the fragment ions at $m/z$ 449 ($[Y_1]^+$), and at $m/z$ 287 ($[Y_0]^+$) in the high-energy function in the positive ion mode. This fragmentation pattern revealed the loss of rhamnosyl group followed by a hexosyl group, which is in agreement with the fragment ions observed in the negative ion mode, i.e., $[Y_1]^-$ at $m/z$ 447 and $[Y_0]^-$ at $m/z$ 285. In the negative ion mode, both compounds yielded the deprotronated molecule as the base peak in both low and high-energy functions, supporting their tentatively assignment as luteolin-rhamnosylhexoside. Compound **85** was tentatively identified as luteolin-7-*O*-rutinoside since it was the major compound and has been previously found in different lettuce cultivars [18]. The second luteolin-rhamnosylhexoside (**84**) is here reported for the first time in lettuce to the authors' knowledge.

Compound **86** (Rt = 11.85 min, λmax = 265, 339 nm) yielded in the positive mode the protonated molecule at $m/z$ 553, $[M+Na]^+$ at $m/z$ 575, $[2(M–H_2O)+H]^+$ at $m/z$ 1069, $[M–H_2O+H]^+$ at $m/z$ 535, $[M+H–$ hydroxymalonyl]$^+$ at $m/z$ 449 and $[Y_0]^+$ at $m/z$ 287; and in the negative mode, $[M–H]^-$ at $m/z$ 551, $[2M–H]^-$ at $m/z$ 1103, $[Y_0]^-$ at $m/z$ 285, and the base peak $[M–H–CO_2]^-$ at $m/z$ 507, which confirmed the presence of the malonyl moiety in the molecule. Furthermore, luteolin characteristic minor product ions at $m/z$ 199 and 175 were present in the negative high-energy spectrum. Therefore, compound **86** was tentatively identified as luteolin-hydroxymalonylhexoside, which has not been previously reported in lettuce in literature as far as we are aware.

Regarding apigenin derivatives, the observation of neutral losses of the conjugated groups and the product ions at $m/z$ 271 and 269 in the positive and negative ion modes, respectively, indicated the presence of apigenin in their structure. Thus, compound **87** (Rt = 20.57 min) showing a loss of 176 $u$ was identified as apigenin-glucuronide; compound **88** (Rt = 23.02 min, λmax = 259, 328 nm) with a loss of 162 $u$, as apigenin-glucoside; and compound **89** (Rt = 23.90 min) with subsequent losses of 146 $u$ and 162 $u$, as apigenin-rhamnosylhexoside, which is here reported for the first time in lettuce cultivars. Compound **90** (Rt = 14.92 min) yielded in the positive ion mode the protonated molecule at $m/z$ 465, $[M+Na]^+$ at $m/z$ 487, $[Y_0]^+$ at $m/z$ 271 and several minor fragments at $m/z$ 163, 153, 145, 121 and 91 that contributed to confirm that the aglycone was apigenin [7, 54]. Accordingly, in the negative ion mode, it produced the deprotonated molecule at $m/z$ 463 and $[Y_0]^-$ at $m/z$ 269. The loss of 194.0427 $u$ ($C_6H_{10}O_7$) observed was tentatively assigned to a pentahydroxyhexanoic residue. Likewise, compound **91** (Rt = 26.99 min) yielded the protonated and deprotonated molecules at $m/z$ 839 and 837 and the corresponding apigenin aglycone ions in positive and negative ion modes, respectively, showing a monoisotopic loss of 568.2731 $u$ ($C_{25}H_{44}O_{14}$), however, its identity was not able to be disclosed with the available spectral data. Apigenin-glucuronide

(**87**) was detected only in the butterhead lettuce cultivar and in the green oak-leaf variety, but not in the red cultivar. Apigenin-glucuronide (**87**) and apigenin-glucoside (**88**) have been already found in lettuce [15, 23]. Found an apigenin-*O*-derivative with the same fragmentation pattern as apigenin-rhamnosylhexoside (**89**) in different lettuce cultivars, as well as luteolin aglycone (**92**, Rt = 27.08 min). However, apigenin-pentahydroxy-hexanoide (**90**), only detected in the red oak-leaf cultivar, and the apigenin conjugate (**91**) have not been previously reported.

### 7.3.2.3 FLAVANONES

A flavanone glycoside was detected and identified on the basis of its UV-visible spectrum and mass spectral data. Chromatographic peak **93** (Rt = 14.87 min, λmax = 284 nm, shoulder at 329 nm) in the negative mode yielded the base peaks [M–H]⁻ at *m/z* 463 in the low energy function, and a fragment ion [$^{1,3}$A]⁻ at *m/z* 151 and an intense ion [$Y_0$]⁻ at *m/z* 287 (60% RA) in the high energy function. In the positive ion mode, [M+H]⁺ at *m/z* 465 (60% RA), [M+Na]⁺ at *m/z* 487 and a fragment ion [$Y_0$]⁺ at *m/z* 289 (base peak) were detected (Figure 7.3). Both fragment ions revealed the cleavage of a glucuronic group. Moreover, minor fragments in the positive ion mode at *m/z* 153, 135, and 117 contributed to confirm that the aglycone was eriodictyol [7]. Thus, compound **93** was identified as eriodictyol-*O*-glucuronide, which is reported for the first time in lettuce to our best knowledge.

### 7.3.2.4 ANTHOCYANINS

Four anthocyanins (**94–97**) were detected in red oak lettuce leaves, despite not working in the optimal pH conditions for their chromatographic separation. Anthocyanidins are ionized much better in the positive ion mode, producing ions with higher intensities in this mode. Therefore, the ions of compounds in very low concentrations (**94** and **97**) were not detected in the negative ion mode.

Compound **94** (Rt = 10.80 min) was identified as a hexoside of cyanidin on the basis of its mass spectra with a [M]⁺ at *m/z* 449, which yielded a high energy function fragment at *m/z* 287. The loss of 162 *u* indicated cleavage

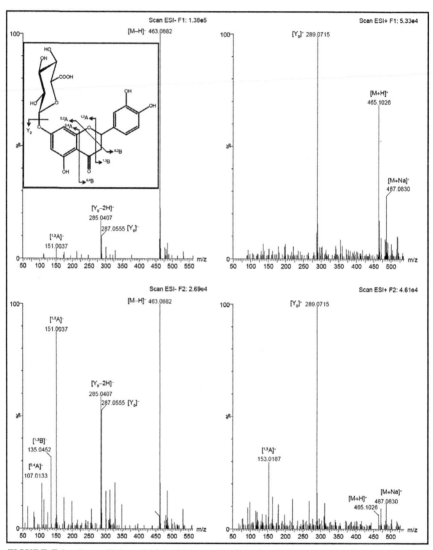

**FIGURE 7.3** Low (F1) and high (F2) energy function MS spectra in the negative and positive ion mode of eriodictyol-*O*-glucuronide (ESI: electrospray ionization).

of a hexosyl residue. Its identification was confirmed by comparison with cyanidin-3-*O*-glucoside standard and bibliographic references [35].

The MS spectra of compounds **95** (Rt = 13.62 min) and **96** (Rt = 16.84 min, λmax = 279, >500 nm) showed their base peaks [M]⁺ at *m/z* 535 and the main product ion $[Y_0]^+$ at $m/z$ 287, which corresponded to cyanidin aglycone and disclosed the loss of a malonylhexosyl residue (Figure 7.4). Both compounds presented the same fragmentation pattern, but regarding the elution order for cyanidin-glycosides on non-end-capped C18 packings observed [35], compound **95** was tentatively identified as cyanidin-3-*O*-(3"-*O*-malonyl)glucoside, and compound **96** as cyanidin-3-*O*-(6"-*O*-malonyl) glucoside. The latter has been reported to be the most abundant anthocyanin in red leaf lettuce varieties, as well as observed here [16, 35].

Another cyanidin glycoside eluting at 20.25 min (**97**) presented a protonated molecule at $m/z$ 491 in the positive ion mode. The base peak in the low energy function was yielded at $m/z$ 449, which revealed the loss of an acetyl residue; and in the high-energy function, the cyanidin aglycone ion at $m/z$ 287. Regarding these observations and bibliographic data [35], compound **97** was tentatively identified as cyanidin-3-*O*-(6"-*O*-acetyl) glucoside, which has been previously found in red leaf lettuce. Cyanidin-3-*O*-glucoside (**94**) and cyanidin-3-*O*-(6"-*O*-malonyl)glucoside (**96**) have been determined in the red oak leaf cultivar before; however, cyanidin-3-*O*-(3"-*O*-malonyl)glucoside (**95**) and cyanidin-3-*O*-(6"-*O*-acetyl)glucoside (**97**) are here reported for the first time in this red cultivar [16, 32, 33].

### 7.3.3  COUMARINS

Seven coumarins (**98–104**) were detected in the three lettuce cultivars studied. Chromatographic peak **98** (Rt = 6.50 min, λmax = 290, 340 nm) was identified as a 6,7-dihydroxycoumarin-6-*O*-glucoside (esculin) regarding its UV-visible spectrum and mass spectral data. In the positive ion mode, the protonated molecule at $m/z$ 341, the sodium adduct at $m/z$ 363 and $[Y_0]^+$ at $m/z$ 179 were produced, indicating that a hexosyl group was present in the molecular structure. This was confirmed in the negative ion mode, where the deprotonated molecular at $m/z$ 339, the acetate adduct $[M–H+AcO]^-$ at $m/z$ 399 and $[Y_0]^-$ at $m/z$ 177 were yielded. Compound **98** also gave some minor fragment ions at $m/z$ 133 and 105 corresponding to the loss of $CO_2$ and CO successively, which have been previously reported in literature, and suggested that peak **98** was esculetin-6-*O*-glucoside [23].

Compounds **99** (Rt = 7.31 min), **100** (Rt = 10.23 min) and **101** (Rt = 12.02 min, λmax = 296, 330 nm) presented the same protonated molecules at $m/z$ 179 and deprotonated molecules at $m/z$ 177, as well as the same frag-mentation pattern described above for esculin. Thus, they were tentatively

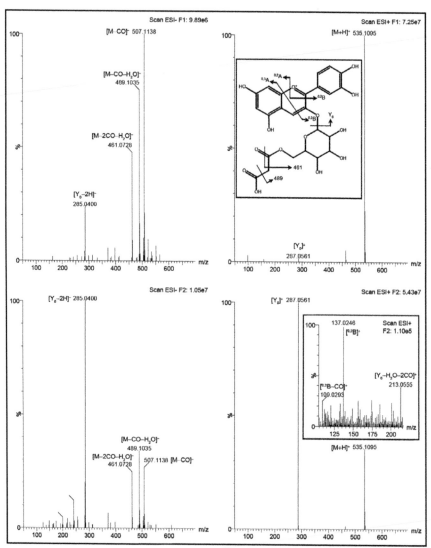

**FIGURE 7.4**  Low (F1) and high (F2) energy function MS spectra in the negative and positive ion mode of cyanidin-3-*O*-(6"-*O*-malonyl)glucoside, ESI, electrospray ionization.

identified as dihydrocoumarin isomers. Esculin and 6,7-dihydrocoumarin (**101**) have been already reported in lettuce and *Asteraceae* [23, 55]. In the same way, compounds **102** (Rt = 9.05 min), **103** (Rt = 10.54 min) and **104** (Rt = 12.54 min) presented the same fragmentation patterns as the dihydrocoumarin isomers, but their protonated molecules at $m/z$ 295 and deprotonated molecules at $m/z$ 293 disclosed that the loss to yield the dihydrocoumarin ion was 116 $u$, due to a maloyl residue. Thus, these compounds were tentatively assigned as maloyl-dihydrocoumarin isomers. Regarding the elution order of the dihydrocoumarin and the maloyl-dihydrocoumarin isomers, the latters are probably the maloyl derivatives of the formers, since the maloyl group increase the hydrophobicity of the molecule, and therefore, elute at higher RTs in reverse-phase packings. To the authors' knowledge, maloyl-dihydrocoumarins are reported in lettuce and *Asteracea* for the first time and all these coumarins are also here described for the first time in green and red oak-leaf lettuce cultivars.

### 7.3.4 HYDROLYSABLE TANNINS

A tri-4-hydroxyphenylacetyl ester of a hexose (**105**, Rt = 27.09 min) was detected in the extracted trace at $m/z$ 581 in the negative ion mode. This peak showed the characteristic fragmentation pattern previously described in literature, yielding fragment ions at $m/z$ 295 ([(4-hydroxyphenylacetic acid-hexose)–H–$H_2O$]⁻), $m/z$ 175 ([(4-hydroxyphenylacetic acid-hexose)–2H–$H_2O$–$C_6H_5CH_2CO$]⁻), $m/z$ 151 ([4-hydroxyphenylacetic acid–H]⁻ and $m/z$ 143 ([[(4-hydroxyphenylacetic acid-hexose)–2H–$H_2O$–$OHC_6H_4CH_2COOH$]⁻ or [hexose–H–$2H_2O$]⁻) [23, 56]. Four isomers of tri-4-hydroxyphenylacetyl-glucoside were found in several *Latuca* species, but is here reported for the first time in green and red oak-leaf lettuce cultivars as far as we know [23, 56].

### 7.3.5 LIGNAN DERIVATIVES

Peak **106** (Rt = 21.00 min), detected in the extracted MS chromatogram set at $m/z$ 417 in the negative ion mode only in the butterhead lettuce variety, yielded the fragment ion $m/z$ 359 due to the losses of two methyl moieties plus CO. In the positive ion mode, the corresponding protonated molecule was detected at $m/z$ 419. This compound was tentatively identified as syringaresinol, having not been found in lettuce cultivars before to the best of our knowledge. In relation to this compound, four syringaresinol-hexoses (**107**, Rt = 13.90 min; **108**, Rt = 18.97 min; **109**, Rt = 19.63 min; **110**, Rt = 23.30 min) were detected in the extracted trace at $m/z$ 579 and 581 in the negative and positive ion modes in the three lettuce cultivars studied. For peak **108**, only the corresponding deprotonated and protonated molecules were detected due to its low concentration in the extract. All other isomers yielded in the negative ion mode the fragment ions corresponding to the loss of the hexose residue ($m/z$ 417), and the subsequent losses of $H_2O$ ($m/z$ 399) or two methyl residues ($m/z$ 387) from the syringaresinol. In the positive ion mode, the sodium adducts ($m/z$ 603) and the fragment ion due to the loss of the hexose residue plus two $H_2O$ ($m/z$ 383) were detected. In addition, three isomers of syringaresinol-acetylhexoses (**111**, Rt = 15.06 min, λmax = 205, 280 nm; **112**, Rt = 24.50 min; **113**, Rt = 24.63 min) were detected in the extracted trace at $m/z$ 621 in the negative ion mode, presenting the same aforementioned fragmentation pattern. In this sense, the fragment ions due to the loss of the acetylhexose residue ($m/z$ 417), and the successive losses of $H_2O$ ($m/z$ 399), and methyl residues ($m/z$ 402 (–$CH_3$), $m/z$ 387 (–2$CH_3$)) and $m/z$ 359 (–2$CH_3CO$)) were observed, as well as other further fragments from the syringaresinol structure at $m/z$ 181, 166, 151 and 123.

Peaks **114** (Rt = 19.22 min), **115** (Rt = 19.39 min) and **116** (Rt = 19.82 min) were observed in the chromatogram set at $m/z$ 581 in the negative ion mode. The MS spectra of these compounds disclosed that they presented the same fragmentation pattern as the above lignans, yielding the product ions due to the loss of the dimethoxyhexose moiety ($m/z$ 359), and the subsequent losses of $H_2O$ ($m/z$ 341), and two methyl residues ($m/z$ 329) from the lariciresinol structure. Thus, these compounds were proposed to be isomers of dimethoxy-hexosyl-lariciresinol. Furthermore, a dimethoxy-dihexosyl-lariciresinol isomer (**117**: Rt = 16.37 min) was also tentatively identified according to the presence of the deprotonated ion at $m/z$ 743 and the fragment ion due to the loss of a hexose residue at $m/z$ 581 in its negative ion MS spectra, which yielded further product ions following the same fragmentation pattern of dimethoxy-hexosyl-lariciresinol. In lettuce cultivars, only one isomer of syringaresinol-hexose (syringaresinol-β-D-glucoside) and dimethoxy-hexosyl-lariciresinol have been previously reported [23, 56]. To the authors' knowledge, lignan derivatives are reported for the first time in oak-leaf lettuce cultivars in the present study.

## 7.4   CONCLUSION

High-resolution mass spectrometry plays a very important and relevant role for the structural elucidation of phenolic compounds. In this Chapter, the potential of the analytical methodology based on UHPLC coupled online to diode array detection, ESI, and QToF mass spectrometry (UHPLC-DAD-ESI-QToF/MS), using the automatic and simultaneous acquisition of exact mass at high and low collision energy, i.e., the MS$^E$ acquisition mode, for the characterization of phenolic profiles was proved. As a result, 111 phenolic compounds were tentatively identified in the butterhead lettuce cultivar, 109 in the green oak leaf lettuce cultivar, and 113 compounds in the red cultivar. To the authors' knowledge, the present work reports 48 phenolics not previously reported in lettuce and the first data on the phenolic profile of the butterhead cultivar. By providing the new structural information, UHPLC-DAD-ESI-QToF/MS$^E$ approach allowed the identification of unknown phenolics, demonstrating to be a useful tool for the characterization of phenolic compounds in complex plant matrices.

## ACKNOWLEDGMENTS

The authors gratefully acknowledge the Agencia Nacional dePromoción Científica y Tecnológica (project number PICT-2008-1724) and the Consejo Nacional de Investigaciones Científicas y Técnicas (CONICET) (project number PIP 0007) from Argentina for the financial support. Gabriela Elena Viacava thanks CONICET and Asociación Universitaria Iberoamericana de Postgrado (AUIP) for her PhD grants. Technical and staff support provided by SGIker (UPV/EHU, MICINN, GV/EJ, ESF) is gratefully acknowledged.

## KEYWORDS

- **Lactuca sativa**
- **lettuce**
- **MSE acquisition**
- **phenolic compounds**
- **quadrupole-time-of-flightmass analyzer**
- **time-of-flight mass spectrometry**
- **ultrahigh performance liquid chromatography**

# REFERENCES

1. Fang, C., Fernie, A. R., & Luo, J., (2019). Exploring the diversity of plant metabolism. *Trends Plant Sci., 24*(1), 83–98.
2. Kallscheuer, N., Classen, T., Drepper, T., & Marienhagen, J., (2019). Production of plant metabolites with applications in the food industry using engineered microorganisms. *Curr. Opin. Biotechnol., 56*, 7–17.
3. Ramakrishna, A., & Ravishankar, G. A., (2011). Influence of abiotic stress signals on secondary metabolites in plants. *Plant Signal. Behav., 6*(11), 1720–1731.
4. Dai, J., & Mumper, R. J., (2010). Plant phenolics: Extraction, analysis and their antioxidant and anticancer properties. *Molecules, 15*(10), 7313–7352.
5. Watson, R. R., Preedy, V. R., & Zibadi, S., (2014). *Polyphenols in Human Health and Disease.* Academic Press, Amsterdam.
6. Manach, C., Scalbert, A., Morand, C., Rémésy, C., & Jiménez, L., (2004). Polyphenols: Food sources and bioavailability. *Am. J. Clin. Nutr., 79*(5), 727–747.
7. Abad-García, B., Berrueta, L. A., Garmón-Lobato, S., Gallo, B., & Vicente, F., (2009). A general analytical strategy for the characterization of phenolic compounds in fruit juices by high-performance liquid chromatography with diode array detection coupled to electrospray ionization and triple quadrupole mass spectrometry. *J. Chromatogr. A., 1216*(28), 5398–5415.
8. Allende, A., Martínez, B., Selma, V., Gil, M. I., Suárez, J. E., & Rodríguez, A., (2007). Growth and bacteriocin production by lactic acid bacteria in vegetable broth and their effectiveness at reducing listeria monocytogenes in vitro and in fresh-cut lettuce. *Food Microbiol., 24*(7/8), 759–766.
9. Hung, H. C., Joshipura, K. J., Jiang, R., Hu, F. B., Hunter, D., Smith-Warner, S. A., Colditz, G. A., et al., (2004). Fruit and vegetable intake and risk of major chronic disease. *J. Natl. Cancer Inst., 96*(21), 1577–1584.
10. Soerjomataram, I., Oomen, D., Lemmens, V., Oenema, A., Benetou, V., Trichopoulou, A., Coebergh, J. W., et al., (2010). Increased consumption of fruit and vegetables and future cancer incidence in selected European countries. *Eur. J. Cancer, 46*(14), 2563–2580.
11. Wang, S., Melnyk, J. P., Tsao, R., & Marcone, M. F., (2011). How natural dietary antioxidants in fruits, vegetables and legumes promote vascular health. *Food Res. Int., 44*(1), 14–22.
12. Nicolle, C., Cardinault, N., Gueux, E., Jaffrelo, L., Rock, E., Mazur, A., Amouroux, P., & Rémésy, C., (2004). Health effect of vegetable-based diet: Lettuce consumption improves cholesterol metabolism and antioxidant status in the rat. *Clin. Nutr., 23*(4), 605–614.
13. Serafini, M., Bugianesi, R., Salucci, M., Azzini, E., Raguzzini, A., & Maiani, G., (2002). Effect of acute ingestion of fresh and stored lettuce (*Lactuca sativa*) on plasma total antioxidant capacity and antioxidant levels in human subjects. *Br. J. Nutr., 88*(6), 615–623.
14. DuPont, M. S., Mondin, Z., Williamson, G., & Price, K. R., (2000). Effect of variety, processing, and storage on the flavonoid glycoside content and composition of lettuce and endive. *J. Agric. Food Chem., 48*(9), 3957–3964.

15. Alarcón-Flores, M. I., Romero-González, R., Martínez, V. J. L., & Garrido, F. A., (2016). Multiclass determination of phenolic compounds in different varieties of tomato and lettuce by ultra-high-performance liquid chromatography coupled to tandem mass spectrometry. *Int. J. Food Prop., 19*(3), 494–507.

16. Marin, A., Ferreres, F., Barberá, G. G., & Gil, M. I., (2015). Weather variability influences color and phenolic content of pigmented baby leaf lettuces throughout the season. *J. Agric. Food Chem., 63*(6), 1673–1681.

17. Pepe, G., Sommella, E., Manfra, M., De Nisco, M., Tenore, G. C., Scopa, A., Sofo, A., et al., (2015). Evaluation of anti-inflammatory activity and fast UHPLC-DAD-IT-TOF profiling of polyphenolic compounds extracted from green lettuce (*Lactuca sativa* L.; var. *Maravilla de Verano*). *Food Chem., 167*, 153–161.

18. Llorach, R., Martínez-Sánchez, A., Tomás-Barberán, F. A., Gil, M. I., & Ferreres, F., (2008). Characterization of polyphenols and antioxidant properties of five lettuce varieties and escarole. *Food Chem., 108*(3), 1028–1038.

19. Oh, M. M., Carey, E. E., & Rajashekar, C., (2009). Environmental stresses induce health-promoting phytochemicals in lettuce. *Plant Physiol. Biochem., 47*(7), 578–583.

20. Romani, A., Pinelli, P., Galardi, C., Sani, G., Cimato, A., & Heimler, D., (2002). Polyphenols in greenhouse and open-air-grown lettuce. *Food Chem., 79*(3), 337–342.

21. Ignat, I., Volf, I., & Popa, V. I., (2011). A critical review of methods for characterization of polyphenolic compounds in fruits and vegetables. *Food Chem., 126*(4), 1821–1835.

22. Eugster, P. J., Guillarme, D., Rudaz, S., Veuthey, J. L., Carrupt, P. A., & Wolfender, J. L., (2011). Ultra-high pressure liquid chromatography for crude plant extracts profiling. *J. AOAC Int., 94*(1), 51–70.

23. Abu-Reidah, I., Contreras, M., Arráez-Román, D., Segura-Carretero, A., & Fernández-Gutiérrez, A., (2013). Reversed-phase ultra-high-performance liquid chromatography coupled to electrospray ionization-quadrupole-time-of-flight mass spectrometry as a powerful tool for metabolic profiling of vegetables: *Lactuca sativa* as an example of its application. *J. Chromatogr. A, 1313*, 212–227.

24. Rogachev, I., & Aharoni, A., (2011). UPLC-MS-based metabolite analysis in tomato. In: Hardy, N., & Hall, R., (eds.), *Plant Metabolomics* (pp. 129–144). Humana Press: New York.

25. Ramirez-Ambrosi, M., Abad-Garcia, B., Viloria-Bernal, M., Garmon-Lobato, S., Berrueta, L., & Gallo, B., (2013). A new ultrahigh performance liquid chromatography with diode array detection coupled to electrospray ionization and quadrupole time-of-flight mass spectrometry analytical strategy for fast analysis and improved characterization of phenolic compounds in apple products. *J. Chromatogr. A, 1316*, 78–91.

26. Markham, K. R., (1982). *Techniques of Flavonoid Identification*. Academic Press Inc. London.

27. Ma, Y. L., Li, Q. M., Van, D. H. H., & Claeys, M., (1997). Characterization of flavone and flavonol aglycones by collision-induced dissociation tandem mass spectrometry. *Rapid Commun. Mass Spectrom., 11*(12), 1357–1364.

28. Lozac'h, N., (1975). Nomenclature of cyclitols. *EuropeanJ. Biochem., 57*, 1–7.

29. Viacava, G. E., Roura, S. I., Berrueta, L. A., Iriondo, C., Gallo, B., & Alonso-Salces, R. M., (2017). Characterization of phenolic compounds in green and red oak-leaf

lettuce cultivars by UHPLC-DAD-ESI-QToF/MS using MSE scan mode. *J. Mass Spectrom., 52*(12), 873–902.

30. Viacava, G. E., Roura, S. I., López-Márquez, D. M., Berrueta, L. A., Gallo, B., & Alonso-Salces, R. M., (2018). Polyphenolic profile of butterhead lettuce cultivar by ultrahigh performance liquid chromatography coupled online to UV-visible spectrophotometry and quadrupole time-of-flight mass spectrometry. *Food Chem., 260*, 239–273.

31. Sofo, A., Lundegårdh, B., Mårtensson, A., Manfra, M., Pepe, G., Sommella, E., De Nisco, M., et al., (2016). Different agronomic and fertilization systems affect polyphenolic profile, antioxidant capacity, and mineral composition of lettuce. *Sci. Hortic., 204*, 106–115.

32. Becker, C., Klaering, H. P., Kroh, L. W., & Krumbein, A., (2014). Cool-cultivated red leaf lettuce accumulates cyanidin-3-O-(6″-O-malonyl)-glucoside and caffeoylmalic acid. *Food Chem., 146*, 404–411.

33. Becker, C., Klaering, H. P., Schreiner, M., Kroh, L. W., & Krumbein, A., (2014). Unlike quercetin glycosides, cyanidin glycoside in red leaf lettuce responds more sensitively to increasing low radiation intensity before than after head formation has started. *J. Agric. Food Chem., 62*(29), 6911–6917.

34. Santos, J., Oliveira, M., Ibáñez, E., & Herrero, M., (2014). Phenolic profile evolution of different ready-to-eat baby-leaf vegetables during storage. *J. Chromatogr. A., 1327*, 118–131.

35. Wu, X., & Prior, R. L., (2005). Identification and characterization of anthocyanins by high-performance liquid chromatography-electrospray ionization-tandem mass spectrometry in common foods in the United States: Vegetables, nuts, and grains. *J. Agric. Food Chem., 53*(8), 3101–3113.

36. Gómez-Romero, M., Segura-Carretero, A., & Fernández-Gutiérrez, A., (2010). Metabolite profiling and quantification of phenolic compounds in methanol extracts of tomato fruit. *Phytochemistry, 71*(16), 1848–1864.

37. Alonso-Salces, R. M., Guillou, C., & Berrueta, L. A., (2009). Liquid chromatography coupled with ultraviolet absorbance detection, electrospray ionization, collision-induced dissociation, and tandem mass spectrometry on a triple quadrupole for the online characterization of polyphenols and methylxanthines in green coffee beans. *Rapid Commun. Mass Spectrom., 23*(3), 363–383.

38. Clifford, M. N., Johnston, K. L., Knight, S., & Kuhnert, N., (2003). Hierarchical scheme for LC-MSⁿidentification of chlorogenic acids. *J. Agric. Food Chem., 51*(10), 2900–2911.

39. Clifford, M. N., Knight, S., & Kuhnert, N., (2005). Discriminating between the six isomers of dicaffeoylquinic acid by LC-MSⁿ. *J. Agric. Food Chem., 53*(10), 3821–3832.

40. Clifford, M. N., Knight, S., Surucu, B., & Kuhnert, N., (2006). Characterization by LC-MSn of four new classes of chlorogenic acids in green coffee beans: Dimethoxycinnamoylquinic acids, diferuloylquinic acids, caffeoyl-dimethoxcinnamoylquinic acids, and feruloyl-dimethoxycinnamoylquinic acids. *J. Agric. Food Chem., 54*(6), 1957–1969.

41. Clifford, M. N., Wu, W., Kirkpatrick, J., & Kuhnert, N., (2007). Profiling the chlorogenic acids and other caffeic acid derivatives of herbal chrysanthemum by LC-MS$^n$. *J. Agric. Food Chem.*, *55*(3), 929–936.

42. Jaiswal, R., Kiprotich, J., & Kuhnert, N., (2011). Determination of the hydroxycinnamate profile of 12 members of the Asteraceae family. *Phytochemistry*, *72*(8), 781–790.

43. Clifford, M. N., Kirkpatrick, J., Kuhnert, N., Roozendaal, H., & Salgado, P. R., (2008). LC-MSn analysis of the cis isomers of chlorogenic acids. *Food Chem.*, *106*(1), 379–385.

44. Dawidowicz, A. L., & Typek, R., (2011). The influence of pH on the thermal stability of 5-O-caffeoylquinic acids in aqueous solutions. *Eur. Food Res. Technol.*, *233*(2), 223–232.

45. Jeong, S. W., Kim, G. S., Lee, W. S., Kim, Y. H., Kang, N. J., Jin, J. S., Lee, G. M., et al., (2015). The effects of different nighttime temperatures and cultivation durations on the polyphenolic contents of lettuce: Application of principal component analysis. *J. Adv. Res.*, *6*(3), 493–499.

46. Ribas-Agustí, A., Gratacós-Cubarsí, M., Sárraga, C., García-Regueiro, J. A., & Castellari, M., (2011). Analysis of eleven phenolic compounds including novel p-Coumaroyl derivatives in lettuce (*Lactuca sativa* L.) by ultra-high-performance liquid chromatography with photodiode array and mass spectrometry detection. *Phytochem. Anal.*, *22*(6), 555–563.

47. Sobolev, A. P., Brosio, E., Gianferri, R., & Segre, A. L., (2005). Metabolic profile of lettuce leaves by high-field NMR spectra. *Magn. Reson. Chem.*, *43*(8), 625–638.

48. Lin, L. Z., Harnly, J., Zhang, R. W., Fan, X. E., & Chen, H. J., (2012). Quantitation of the hydroxycinnamic acid derivatives and the glycosides of flavonols and flavones by UV absorbance after identification by LC-MS. *J. Agric. Food Chem.*, *60*(2), 544–553.

49. Mai, F., & Glomb, M. A., (2013). Isolation of phenolic compounds from iceberg lettuce and impact on enzymatic browning. *J. Agric. Food Chem.*, *61*(11), 2868–2874.

50. Clifford, M. N., Marks, S., Knight, S., & Kuhnert, N., (2006b). Characterization by LC-MSn of four new classes of p-coumaric acid-containing diacyl chlorogenic acids in green coffee beans. *J. Agric. Food Chem.*, *54*(12), 4095–4101.

51. Heimler, D., Isolani, L., Vignolini, P., Tombelli, S., & Romani, A., (2007). Polyphenol content and antioxidative activity in some species of freshly consumed salads. *J. Agric. Food Chem.*, *55*(5), 1724–1729.

52. Gómez-Romero, M., Zurek, G., Schneider, B., Baessmann, C., Segura-Carretero, A., & Fernández-Gutiérrez, A., (2011). Automated identification of phenolics in plant-derived foods by using library search approach. *Food Chem.*, *124* (1), 379–386.

53. Katta, V., Chowdhury, S. K., & Chait, B. T., (1991). Use of a single-quadrupole mass spectrometer for collision-induced dissociation studies of multiply charged peptide ions produced by electrospray ionization. *Anal. Chem.*, *63*(2), 174–178.

54. Abad-García, B., (2007). Búsqueda de marcadores químicos de tipo polifenólico para la autentificación de zumos de frutas. Universidad del País Vasco/Euskal Herriko Unibertsitatea (UPV/EHU), Bilbao (Spain), 865.

55. Schütz, K., Carle, R., & Schieber, A., (2006). Taraxacum: A review on its phytochemical and pharmacological profile. *J. Ethnopharmacol.*, *107*(3), 313–323.

56. Sessa, R. A., Bennett, M. H., Lewis, M. J., Mansfield, J. W., & Beale, M. H., (2000). Metabolite profiling of sesquiterpene lactones from *Lactuca* species major latex components are novel oxalate and sulfate conjugates of lactucin and its derivatives. *J. Biol. Chem., 275*(35), 26877–26884.
57. Winkel-Shirley, B., (2001). Flavonoid biosynthesis: A colorful model for genetics, biochemistry, cell biology, and biotechnology. *Plant Physiol., 126*(2), 485–493.

# CHAPTER 8

# Applications of Gas Chromatography-High-Resolution Mass Spectrometry (GC-HRMS) for Food Analysis

JANET ADEYINKA ADEBIYI,[1] PATRICK BERKA NJOBEH,[1] NOMALI NGOBESE,[2] GBENGA ADEDEJI ADEWUMI,[3] and OLUWAFEMI AYODEJI ADEBO[1]

[1]*Department of Biotechnology and Food Technology, Faculty of Science, University of Johannesburg, P.O. Box – 17011, Doornfontein, Johannesburg, South Africa, E-mails: janetaadex@gmail.com (J. A. Adebiyi), oadebo@uj.ac.za (O. A. Adebo)*

[2]*Department of Botany and Plant Biotechnology, Faculty of Science, University of Johannesburg, P.O. Box – 524, Auckland Park, Johannesburg, South Africa*

[3]*Department of Microbiology, Faculty of Science, University of Lagos, Akoka, Lagos, Nigeria*

## ABSTRACT

With the drive and context for the detection of more compounds and components in foods, the need for a robust, sensitive, and accurate system for the detection of constituents has never been more important. From a wide range of available analytical platforms for foods, gas chromatography-high resolution mass spectrometry (GC-HRMS) provides a better analytical technique, due to its high precision mass resolution, reproducible fragmentation patterns, as well as high scan speeds. This subsequently provides a platform which can give insights, both for targeted forms of analysis (known), as well as untargeted approaches (both known and unknowns). Although, this equipment has been utilized for various analyzes in foods, there is potential for its utilization in understanding the composition of other food components.

## 8.1 INTRODUCTION

It is gradually becoming much more important that precise estimations of food constituents be reported and made available for its teeming consumers. Not only is this vital to mitigate against food fraud and associated issues, such precisions also help in ensuring persistently safe food products. Principally, food analytical techniques involve both subjective and objective (instrumental) determinations to ascertain inherent composition and acceptability of food materials. Such liberated compounds might be due to chemical changes brought about by inclusion of additives and/or unintended chemical reaction of contaminants (e.g., agrochemicals, environmentally induced contaminants, toxins, and leaching of packaging elements) in addition to activities along the food chain [1].

According to Kachlicki et al. [2], mass spectrometry (MS) is a robust tool for structural characterization of compounds. Among other widely used analytical MS-based methods for food analysis, there is a growing interest in the use of gas chromatography-mass spectrometry (GC-MS), which is in part due to its high separation efficiency, as well as sensitive and selective mass detection capabilities [3–5]. Further to this is the relatively easier assignment of peaks through available database, due to reproducible fragmentation patterns obtained at 70 eV [6]. Emerging constituents, adulteration challenges, residues of contaminants, as well as other new insights into foods have however, necessitated a better and improved analytical platform. This is important due to numerous disadvantages of low-resolution GC instruments, particularly the limitation of only selected ion monitoring (SIM) types of analysis [7]. Subsequent developments in this regard have led to the birth of high resolution GC-MS, which has been referred to as gas chromatography-electron ionization-Orbitrap-high resolution accurate mass spectrometry (GC-EI-Orbitrap-HRAMS), gas chromatography-high resolution accurate mass spectrometry (GC-HRAMS), and gas chromatography-high resolution time of flight mass spectrometry (GC-HRToF-MS).

The subsequent adoption of high-resolution mass spectrometry (HRMS) in GC is due to its low mass experimental errors, high mass resolutions and data acquisition rates [8, 9]. High-resolution mass spectrometry has emerged as a powerful tool for non-targeted analysis sequel to its ability to identify compounds and possibly quantify them at trace levels. The technology has seen application in the detection of metabolites in fermented foods and beverages [3, 10], plant materials [11], as well as pesticides in fruits and vegetables [9, 12]. This chapter appraises studies that have applied GC-HRMS and illustrate the applicability of this technique in the study of foods for constituents such as pesticides, polyaromatic hydrocarbons (PAHs) and other contaminants, in addition to inherent metabolites (secondary and primary).

## 8.2  FUNDAMENTALS AND BACKGROUND ON GC-MS

The need to quantitatively and qualitatively distinctively measure and characterize vital properties in relation to quality, consistency, and safety has evolved into several instrumental innovations for analysis. Gas chromatography is a common column chromatography technique. When it is coupled with mass spectrometry, it functions as a hybrid analytical technique that can be used for characterizing volatile, partly volatile, and partly polar chemical compounds [1]. The concept of pairing gas chromatography to mass spectrometry (hyphenation) came into being after the first merchandise of quadrupole mass spectrometer (MS), with an extension of a data processor system. Subsequently, substantial volumes of inherent compounds in sample matrices limited the examination of targeted analytes, which led to the innovation of advanced mass spectrometry with improved selectivity. Such innovation extended to the application of collisionally activated dissociation (CAD) systems, such as tandem mass spectrometry (MS-MS) coupled with triple-stage quadrupole (QqQ), and patented multi-stage mass spectrometry, to enhance selectivity, efficacy, and sensitivity during analysis. In the same vein, just like the preceding mechanism, inadequacies were realized, ranging from partial screening of targeted chemicals, less efficient to detect untargeted compounds to dependence on reference standards. These defects unfold the introduction of more sophisticated tools capable of generating a wider range of spectra data, higher mass accuracy and resolving power, enumeration of unknown compounds and without the use of reference standards. Such significant technological output is the emergence of gas chromatography paired with high-resolution mass spectrometry in food analysis. The twenty-first century opened up the possibilities of advanced high-resolution mass spectrometry instruments; time-of-flight (ToF) and Orbitrap instrumentations, either single, hybrid or tribrid analyzers [13].

## 8.2.1  MODE OF OPERATION OF GAS CHROMATOGRAPHY-MASS SPECTROMETRY (GC-MS)

This class of column chromatography is majorly composed of two phases: mobile (inert gas) and stationary (solid or immobilized polymeric liquid). As such, typical gas chromatography may be categorized as gas-solid chromatography (G-SC) or the frequently used gas-liquid chromatography (G-LC). Briefly, under-regulated temperature, an inert carrier gas stream (most often helium or nitrogen), containing volatilized analyte travels at constant flow mode to the column fixed with stationary phase. Here, individual component of the matrix under examination are screened on the basis of several physico-chemical properties, including relative vapor pressure and solubility in the immobilized liquid confined in stationary phase [14]. On elution, the generated constituents are ionized in the ionization compartment of mass analyzer for identification and quantification. An example of GC-MS is schematically illustrated in Figure 8.1. Various conditions are pre-defined in GC, these include injection mode (split/splitless) (min), flow rate of carrier gas (ml/min), film thickness (m x mm x mm), temperature, among others [3, 12].

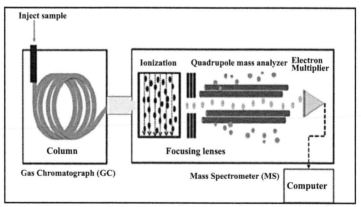

FIGURE 8.1    Schematic diagram of gas chromatography paired with a mass spectrometer.
*Source*: Reprinted from Ref. [15]. http://creativecommons.org/licenses/by/4.0/

## 8.2.2 AN OVERVIEW OF GC-HRMS

Basically, mass spectrometry estimates the mass-to-charge ratio ($m/z$) of organic or inorganic matrices of known charges. $m/z$ is a dimensionless term with respect to IUPAC classification, where $m$ is the relative molecular mass of an ion in Daltons (Da) divided by its number of charges $z$. Accordingly, the graphical representation of mass spectrum is depicted as relative abundance (RA) versus their respective $m/z$ values of the detected ionic matrices. Notwithstanding, the measurement of mass spectrum may either be to elucidate data relating to integer/nominal mass units (low-resolution mass spectrometry (LRMS)) or exact mass units (high-resolution mass spectrometry-HRMS). This classification distinguishes the analytical competency of HRMS to provide wider spectra of untargeted constituents of a matrix with equivalent nominal mass but differs in monoisotopic mass [16]. Considering the importance and role of resolution in structural elucidation, identification, as well as detection of unknown compounds, HRMS provides a possibility of such thorough information about elemental composition and precise empirical information. Other principal factors that highlight the screening capabilities of HRMS analyzers include range of $m/z$, high sensitivity, mass resolving power, mass accuracy, scan speed, automatic gain control, limits of detection/quantification, recovery, repeatability, and linear dynamic range [3, 12]. HRMS has the capability of answering biological, analytical, and other queries that might be posed in studies. These include but are not limited to constituents/composition, concentration (amounts) and structure in a very fast and sensitive manner. An overview of such application in foods is summarized in Section 8.3 of this chapter.

In a typical GC-MS analytical run, analytes are attached to the surface of a column and individual constituents are intermittently eluted, based on their volatility using a temperature gradient [17]. The signal subsequently gives a response proportional to the concentration of the analyte (compound), which is useful for quantification and can be viewed infographically on a chromatogram (Figure 8.2). As such, GC-HRMS is only suited for the detection of volatile and non-polar metabolites. To improve detection derivatization of GC-samples prior to analysis may be required to improve volatility and make them GC-amenable. Most GC-HRMS instruments involve hyphenation with ToF, quadrupole time-of-flight (QToF) and ion trap (IT) Orbitrap or hybrid quadrupole Orbitrap (Q-Orbitrap) systems. These techniques and ionization sources will be briefly discussed in the succeeding section.

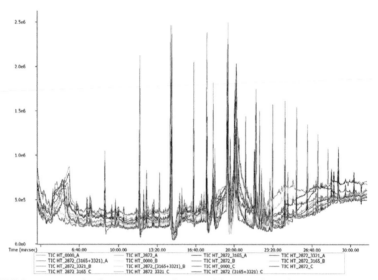

**FIGURE 8.2**    A sample chromatogram of raw and fermented sorghum samples.


## 8.2.2.1   *IONIZATION TECHNIQUES*

The versatility of GC-HRMS can be well reflected in their interchange-able ionization sources and sophisticated data acquisition capabilities. An example of such is the LECO Pegasus HRT-High Resolution GC-MS (Figure 8.3), with capabilities for both electron and chemical ionization (EI and CI).

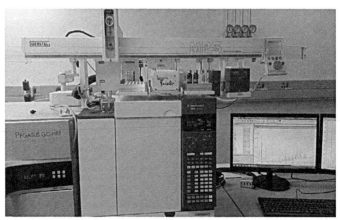

**FIGURE 8.3**    A LECO Pegasus GC-HRTOF-MS instrument with a direct inlet probe at the University of Johannesburg, South Africa

Ionization process in the detection of spectral components is an initial step to liberate ions from analytes in mass spectrometry. To thus allow for adequate and effective analysis of food constituents, the choice of an appropriate ionization is needed during analysis. This is vital knowing that ionization plays a crucial role in the detectability of constituents. It should nonetheless be noted that proper mass calibration of the $m/z$ analyzer is essential for mass accuracy and stability of the mass peaks. Conventionally, either the reference calibrant (also called lock masses) with a set of known $m/z$ values or the automatic use of MS software without the application of mass calibration standard is usually required for the calibration step. As such, MS may be internally or externally calibrated. Various ionization techniques have been discussed in the literature and can be consulted for further reading [18–21]. The ionization methods of significant interest as regards GC-MS are electron and chemical ionization (EI and CI), and these will be briefly highlighted in this chapter.

By means of a heated filament in an ion source (Figure 8.4(A)), an electron beam (generated from a filament or heated tungsten) with kinetic energy (typically 70 eV), is emitted and accelerated to bombard gaseous molecules, and consequently produce positively charged ions and molecules/ions with an unpaired electron in a vacuum [22]. The positively charged fragment ions (written as $M^+$) are drawn out of the ionization compartment by an electric field, electrostatically focused

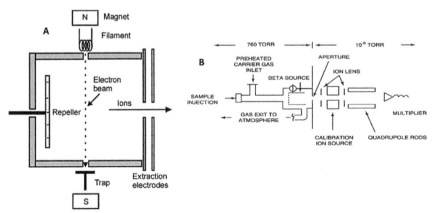

**FIGURE 8.4**   (A) An EI ion source [18]; (B) A CI ion source [23].

*Sources*: A Reprinted with permission from Ref. [18]. © 2019 Elsevier; B: Reprinted with permission from Ref. [23]. © 1973 American Chemical Society.

and path to the mass-to-charge ratio analyzer. Furthermore, fragmentation patterns detected in an EI chamber are subsequently used to authenticate analyte constituent identification. Typical fragmentation pathways include radical elimination to give an even electron ion, a neutral molecule elimination to give another odd-electron ion or fragmentation by rearrangement, often involving migrations of hydrogen atoms [18]. Accordingly, the resulting array of parent and daughter ions (fragments) constitutes what is called the mass spectrum of the sample (Figure 8.5).

For much less polar and small molecules, ambient ionization techniques are more suitable as these yield single charged molecular ions. This has also been reiterated by Harvey [18] that instability of odd-electron ions formed by EI usually results in spectra in which the molecular ion peak is absent as none of the ions survives long enough to reach the detector. These challenges bring CI to the fore as it requires lesser energy and often gives a simpler mass spectrum output with a narrow fragmentation pattern, which improves detection sensitivity. The CI source (Figure 8.4(B)) is constructed in a similar manner to the EI source but has a much smaller exit slit, thus making it much more gastight, allowing pressures of 0.1–1 Torr or less to be achieved [18, 23]. It involves the ionization of a reagent gas by beam of electron and then interacts with gaseous molecules to either transfer or abstract positively charged ions. Compounds such as ammonia, isobutene, methane, or water could be used for the ionization process in CI [20]. Once ionized, these gasses undergo ion-molecule reactions to form species, which then react with sample molecules with a variety of outcomes [18]. As reported by some authors, CI could be positive, involving the following reactions: proton transfer, charge exchange, electrophilic addition, anion extraction or hydride ion extraction or negative with reactions such as electron capture, proton abstraction, charge exchange, and nucleophilic addition [18].

## 8.2.2.2  *MASS ANALYZERS*

For HRMS instruments generally (GC included), ToF mass analyzer appears to be more frequently utilized. ToF-MS provides a platform for capturing a relatively broader molecular weight range of signals

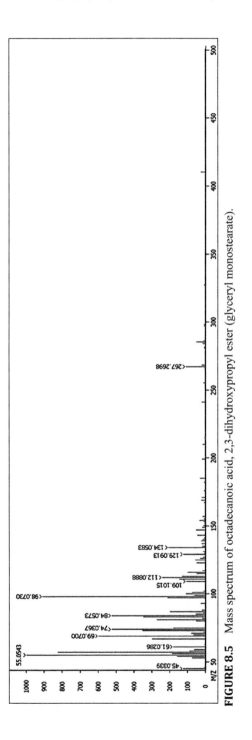

**FIGURE 8.5**   Mass spectrum of octadecanoic acid, 2,3-dihydroxypropyl ester (glyceryl monostearate).

associated with compounds in a single sample/analyte. Ions are produced by bombardment of the sample with particles of MeV energies, usually fission fragments, and the pulses are formed by individual bombarding particles arriving at the target [24]. A parallel beam of ions emitted by an ionization source (EI or CI) travels into the ion pulser through the ion optics and focusing region. Ion's flight to the detector is therefore triggered with the application of a high voltage pulse to the pusher plate, to orthogonally accelerate pockets of ions, based on the velocity of *m/z* into the ToF flight tube and is subsequently detected [25]. According to Stewart et al. [26], advantages of using ToF for broad-spectrum analysis include increased mass accuracy and mass resolution, greater sensitivity, rapid acquisition, and increased dynamic range when profiling over a broad molecular weight range.

Another frequently used mass analyzer for GC is the Orbitrap. According to Perry et al. [27], the Orbitrap uses only electrostatic fields to confine and to analyze injected ion populations. It has been available for several years on liquid chromatography mass spectrometry (LC-MS) platforms, but more recently available on GC-MS platforms [14]. An analytical instrumentation of orbital ion confinement is composed of coupled axisymmetrical electrodes *viz.* the central coaxial spindled-shaped electrode and the barrel-like outer electrode, with an electrostatic field operating between the inner and outer electrodes. Charged ion packets built up on a foreign injecting mechanism (C-trap) are tangentially injected through a slit in the outer electrode, trapped by voltage ramp applied to the inner ring electrode, and radially accelerate in a spiral and oscillating linear motion along the central electrode. By increasing the effect of *mass-to-charge* ratio of the ions, they are detected by the amplifier and digitally processed into a spectrum. As with ToF analyzers, the Orbitraps can exist as MS-only systems or as hybrids, for example, with a quadrupole mass filter to perform mass selection prior to analysis of the fragments by the Orbitrap analyzer [28]. In addition to these aforementioned analyzers are quadrupole (Q) HRToF platforms and magnetic sector, both of which have found applications in high-resolution gas chromatography [29]. A summary of these commercially available GC-HRMS instruments has been provided in Table 8.1. The following studies can be consulted for further reading: [25, 28–33].

## 8.3   USE OF HIGH-RESOLUTION GAS CHROMATOGRAPHY/MASS SPECTROMETRY FOR FOOD ANALYSIS

### 8.3.1   PROFILING

There is indeed a vast diversity of metabolites and components in food and the plant kingdom in general. Such constituents are responsible for a number of food qualities, including but not limited to taste, safety, aroma, health-promoting properties, and overall composition. Profiling is thus primarily aimed at providing an overview and detecting as many metabolites as possible in a food sample at a particular point in time. According to a recent study, over 200,000 primary and secondary metabolites may be present in the plant kingdom [34]. Profiling is thus more qualitative and does not necessarily quantify metabolites/provide absolute concentrations. Using a GC-HRToF-MS system, metabolites in naturally and lactic acid bacteria (LAB) fermented *ting* (a Southern African food) with different tannin contents were investigated, using a non-targeted metabolomics approach [3]. The analytical platform gave synchronic quantitation of diverse analytes and resultantly identified significant metabolites responsible for differences in whole grain sorghum and subsequently obtained *ting*.

Given the plethora of volatile metabolites found in fermented products, high-resolution gas chromatography equipped with high specificity of electron impact MS was used to quantify and profile varying end products [10]. The chromatographic analysis was achieved in less than 5 min and over 40 marked and unmarked volatile constituents were detected and identified. The technique also demonstrated low technical variability (<3%), excellent accuracy (100 ± 5%), recovery (100 ± 10%), reproducibility, and repeatability (coefficient of variation 1–10%) illustrating the methods' rapid quantitation rate [10]. The recent study of Hammann et al. [35] also reported the use of gas chromatography-quadrupole-time of flight high-resolution mass spectrometry (GC-Q-ToF-HRMS) for lipid profiling of seven cereals. In combination with multivariate data analysis, the authors separated the cereals using principal component analysis (PCA). The authors identified the approach as a novel one for lipid profiling, based on high temperature-GC, and also sensitive enough for the detection of analytical and nutritionally important compounds. Other similar studies on profiling foods using high resolution GC-MS are presented in Table 8.2.

**TABLE 8.1** Summary of the Most Important Parameters of Commercially Available GC-HRMS Instruments

| Company | Agilent | Waters | | Leco | Thermo Fisher | | Joel | | Tofwerk |
|---|---|---|---|---|---|---|---|---|---|
| Analyzator type | Q-TOF | Q-TOF | Magnetic sector | TOF | Orbitrap | Magnetic sector | Magnetic sector | TOF | TOF |
| Ionization | EI, PCI, NCI | CI | EI, CI | EI, PCI | EI, CI | EI, CI | EI | EI, CI | EI |
| Mass resolution | >13 500 | >22 500 | >80 000 | >50 000 | >120 000 | >60 000 | >80 000 | >8 000 | >7 000 |
| Linear dynamic range | >$10^5$ | >$10^4$ | >$10^5$ | >$10^3$ | >$10^6$ | >$10^5$ | n.f. | >$10^4$ | >$10^7$ |
| Mass accuracy | <3 ppm RMS | <1 ppm RMS | <5 ppm RMS | <1 ppm RMS | <1 ppm RMS | <2 ppm RMS | n.f. | <4 ppm RMS | <4 ppm |
| Mass range | TOF: 50–1 700 Quadrupol: 50–1 050 | TOF: 20–100 000 Quadrupol: 20–16 000 | n.f. | 10–1 500 | 30–30 000 | 2–6 000 | 1–12 000 | 4–5 000 | >4 000 |
| Instrument detection limit | 1 pg OFN | n.f. | 100 fg TCDD | 1 pg HCB | 6 fg OFN | 20 fg TCDD | n.f. | 1 pg OFN | n.f. |

*Abbreviations:* CI: chemical ionization; EI: electron ionization; fg: femtogram; HCB: hexachlorobenzene; NCI: negative chemical ionization; n.f.: not found; OFN: octafluoronaphtalene; PCI: positive chemical ionization; pg: picogram; ppm: parts per million; RMS: root mean square error; Q-TOF: quadrupole time of flight; TCDD: tetrachlorodibenzo-p-dioxin [29].

**TABLE 8.2**     Some Reported Studies on the Use of GC-HR-MS for Food Analysis

| Foods | Extraction Techniques | Instruments Used | Injection Modes | References |
|---|---|---|---|---|
| | | **Profiling** | | |
| Apple | Liquid/SDE | GC-HRMS/ GC-C/P-HRIRMS | Split | [36] |
| Bread | SAFE | GC × GC-HRMS | NS | [37] |
| Cereals | Liquid | GC-Q-TOF-HRMS | NS | [35] |
| Dezhou braised chicken | ASE and SAFE | GC × GC-HR-TOFMS | Splitless | [38] |
| Dry-cured hams* | MAE and SAFE | GC × GC-HR-TOFMS | Split | [39] |
| Fermentation end products and fermented beverages | QuEChERS | GC-EI-HR-MS | Split | [10] |
| Lupin flour | SAFE | HRGC-O/GC × GC-HRMS | Splitless | [40] |
| Oils | SPME | GC-HRTOF-MS | Splitless | [41] |
| Passion fruit | Headspace | GC-HRMS | Splitless | [42] |
| Saffron | Liquid | HRGC-FID | Split | [43] |
| Scotch whisky | Liquid and SPME | GC-HRMS | Splitless | [44] |
| Sorghum and fermented product (*ting*) | Liquid | GC-HRTOF-MS | Splitless | [3] |
| | | **Pollutants and Contaminants** | | |
| *Acrylamide* | | | | |
| Food | dSPE | GC-HRTOF-MS | Pulsed splitless | [45] |
| *Chlorinated Paraffin* | | | | |
| Fish (salmon) | ASE | GC/ECNI-Orbitrap-HRMS | Splitless | [46] |
| Meat and meat products | ASE | GC × GC-ECNI-TOFMS | NS | [47] |
| Chloropropanol | | | | |
| Meat | PLE | GC-HRMS | NS | [48] |
| *Dechlorane-Related Compounds and Flame Retardants* | | | | |
| Food | Liquid | GC-HRMS | Splitless | [49, 50] |
| Milk and dairy products, eggs, meat, fish, and cod liver | Liquid | GC-MS-HRMS | Splitless | [51] |
| *Polyaromatic Hydrocarbons* | | | | |
| Cereal products | Liquid and GPC | GC-HRMS | Splitless | [52] |
| Foods | ASE | HRGC-ID-HRMS | NS | [53] |
| Smoked meat products and edible oils | ASE, GPC, SPE | GC-HRMS | Splitless | [54] |
| Swedish smoked meat and fish | SPE | GC-HRMS | Pulsed splitless | [55] |
| *Pesticides* | | | | |
| Fruit, vegetable, and wheat | QuEChERS/ (dSPE) | GC-EI-HR-Orbitrap-MS | Splitless | [12] |
| Tomato, apple, leek, and orange | QuEChERS | GC-EI-HR-Orbitrap-MS | Splitless | [9] |
| Vegetables | QuEChERS | GC-Q-Orbitrap-MS | Split | [56] |

**TABLE 8.2** *(Continued)*

| Foods | Extraction Techniques | Instruments Used | Injection Modes | References |
|---|---|---|---|---|
| *Polychlorinated Dibenzo-p-Dioxins, Dibenzofurans/Furans, Dioxin-like Polychlorinated Biphenyls* | | | | |
| Eggs | Liquid | HRGC-HRMS | NS | [57] |
| Eggs, milk, meat, plants, and animal fat | Liquid | GC-HRMS | Splitless | [49] |
| Food samples | Liquid | GC-HRMS | Splitless | [58] |
| Foods of animal and plant origin | Liquid | HRGC-HRMS | NS | [59] |
| Milk, eggs, pork, fish, and game | Liquid | HRGC-HRMS | Splitless | [60] |
| *Toxophane* | | | | |
| Fish, livestock meat, poultry, eggs, vegetables | Liquid | ID-GC-HRMS | | [61] |
| **Others** | | | | |
| Fatty acid Composition | | | | |
| Rice | Liquid | GC-HRMS-FID | Split | [62] |
| *Food Pathogen* | | | | |
| *Listeria monocytogenes* | N/A | HRPGC/MS | Split | [63] |
| *Oligosaccharides* | | | | |
| Food | Liquid | GC-HRMS | Split | [64] |
| *Odour-Active Compounds* | | | | |
| Meat | SPME | HRGC-O | N/A | [65] |
| *Medicinal/Phenolic/Bioactive Compounds* | | | | |
| *Ginkgo biloba* | Liquid | HR-GC-MS | Split | [66] |
| Hop | Liquid | GC-HRaccTOF-MS | Split | [67] |
| Tomato | Liquid | GC-HRMS-Q-Orbitrap | Splitless | [68] |

*volatile profiling; ASE: accelerated solvent extraction; dSPE: dispersive solid phase extraction; GC × GC/HR-TOFMS: two dimensional gas chromatography with high resolution time of flight mass spectrometry; GC × GC-ECNI-TOFMS: two-dimensional gas chromatography coupled to electron capture negative ionization high-resolution time-of-flight mass spectrometry; GC × GC-HRMS: two-dimensional gas chromatography high-resolution mass spectrometry; GC/ECNI-Orbitrap-HRMS: Gas chromatography electron capture negative ion-Orbitrap-high resolution mass spectrometry; GC-EI-HR-Orbitrap-MS: gas chromatography-electron ionization-high resolution-Orbitrap-mass spectrometry; GC-C/P-HRIRMS: gas chromatography-combustion/pyrolysis-high resolution isotope ratio mass spectrometry; GC-HRaccTOF-MS: gas chromatography high resolution-accurate mass-time of flight mass spectrometry; GC-HRMS: gas chromatography high resolution mass spectrometry; GC-HRMS-FID: gas chromatography high resolution mass spectrometry-flame ionization detector; GC-HRMS-Q-Orbitrap: gas chromatography high resolution mass spectrometry-quadrupole-Orbitrap; GC-HRTOF-MS: gas chromatography high resolution time of flight mass spectrometry; GC-MHRMS: gas chromatography magnetic sector high resolution mass spectrometry; GC-Q-TOF-HRMS: gas chromatography-quadrupole-time of flight high resolution mass spectrometry; GPC: gel permeation chromatography; HRGC-ID-HRMS: high-resolution gas chromatography-isotope dilution-high resolution mass spectrometry; HRGC-HRMS: high-resolution gas chromatography-high resolution mass spectrometry; HRGC-O: high resolution gas chromatography-olfactometry; HRPGC/MS: high resolution pyrolysis gas chromatography/mass spectrometry; MAE: microwave assisted extraction; NS: not specified; PCBs: polychlorinated biphenyls; PLE: pressurized liquid extraction; SAFE: solvent-assisted flavor evaporation; SDE: Simultaneous distillation extraction; SPE: solid phase extraction; QuEChERS: Quick easy cheap effective rugged safe.

## 8.3.2   POLLUTANTS AND CONTAMINANTS

The role of food in the survival of humans and the provision of daily nutritional needs cannot be overemphasized. Food from farm to fork is, however, a multidisciplinary approaches involving a number of stake-holders throughout the food chain, with potential for contamination along these stages. Foods are thus exposed to different forms of contaminants/pollutants (biological/environmental, physical, and chemical) along the production chain. One of such are pesticides, a class of chemical compounds found mostly as residues in foods and often previously used to control pest or other forms of life that could be lethal to cultivated plants and animals. Examples of these compounds found in some fruits and vegetables include atrazine, cypermethrin, fenamidone, among others [9]. Mol et al. [12] demonstrated the screening capabilities of a quantum leap gas chromatography full scan high-resolution mass spectrometry equipped with electron ionization (EI) and Orbitrap mass spectrometry system (GC-EI-HR-Orbitrap-MS), and based on dispersive solid phase extraction (dSPE), using pesticides present in fruits and vegetables as a subject of residue analysis. The study confirmed optimum conditions in full scan mode and with a resolving power of 60,000 FWHM reported limits of detection/quantification (0.5 µg kg$^{-1}$), recovery (70 and 120%), repeatability (RSD <10%), linearity (over the range of ≤ 5–250 µg kg$^{-1}$) and EI-Orbitrap spectra had good match values against EI-quadrupole spectra. In a similar pattern, Uclés et al. [9] utilized GC-EI-HR-Orbitrap-MS, adopting a QuEChERS extraction technique, screened for pesticide residues in fruits and vegetables. The optimum condition was attained at 60,000 FWHM and 210 pesticide residues were detected.

Flame-retardants are gradually becoming of significant concern for food safety. While these substances are used on textiles, plastics, electronics, furniture, and clothes to make them less flammable, they unfortunately leach into the air, environment, soil, and water and enter the food chain, subsequently contaminating food. The applicability of high-resolution GC-MS has been demonstrated as being effective for the analysis of these compounds (Table 8.2). Zacs et al. [51] reported the use of high-resolution GC-MS to explore and identify flame-retardants in food products of animal origin. The multi-residue analytical developed method was successfully applied for the analysis of food samples, with the analyzes revealing ubiquitous presence of polybrominated diphenyl ethers

(PBDEs) and dechlorane-related compounds. The qualitative screening of masses at ultra-low concentration reflects the much-needed efficiency to detect smaller occurring lethal contaminants in foods, and thus position an upgrade than the available conventional chromatographic methods.

Acrylamide is a heat-generated food toxicant and a possible human carcinogen with neurotoxic properties [69]. They are usually found in carbohydrates-rich foods and considering the frequency of consumption and number of heat-processed carbohydrates-rich foods, frequent routine analysis of this compound is vital. In line with this, Dunovska et al. [45] developed a method for the analysis of acrylamide in foods using a GC-HRToF-MS system. The authors proposed a novel purification strategy (dispersive solid phase extraction), that provides a significant reduction of matrix co-extracts and reported trueness and reliability of the generated data on the GC-HRToF MS. Likewise are PAHs, a group of hydrocarbons which may be triggered from food at elevated thermal processes [65]. They are thus found in food, especially those that are grilled, dried, roasted, or smoked during preparation. They are hydrophobic compounds that can also be derived from natural sources such as man-made incomplete combustion, volcanic eruption, and forest fires [11]. This ubiquitous carcinogenic organic compound poses serious health challenges due to massive consumption in foods. Hence, precise analytical output of its occurrence in food is receiving a lot of attention. Rozentale et al. [52] tried the application of GC-HRMS to measure trace levels of PAHs in cereals and bread samples. The concentrations of four priority PAHs varied from 0.22–1.62 µg kg$^{-1}$ and 14% of the samples evaluated were higher than the low maximum permitted level established in the European Union. Exceeded limits (<0.01–19 µg kg$^{-1}$) were also quantified for smoked meat products and edible oils while screening for 16 PAHs using GC-HRMS [54]. Considering the significant sensitivity and selectivity of the techniques employed, these analytical criteria open possibilities for appropriate measures to mitigate the presence of PAHs in food products.

Polychlorinated biphenyls (PCBs)/furans/dioxins are substances that represent a lethal group of compounds with alike structural make-up. They have been reported as persistent organic pollutants (POPs), widely spread in the human environs and consumable edible materials [70]. Taking into account the considerable health effects of these toxic contaminants through dietary exposure, comparison intensive analysis of different food samples using atmospheric pressure gas chromatography equipped with tandem mass spectrometry (APGC-MS/MS) and GC-HRMS was unveiled by ten Dam et al. [58]. The techniques had similar linear dynamic range and ion ratio tolerance was within acceptable range of ±15%. In contrast, APGC-MS/MS showed improved sensitivity, while GC-HRMS gave better relative intermediate precision standard deviation ($S_{RW}$, rel) and selectivity. Hence, GC-HRMS was considered as best choice technique. In another study, GC-HRMS screening of selected food samples for toxic compounds was carried out by Godliauskienė et al. [49]. Varying sub-picogram concentration level of POPs congeners [dioxin-like polychlorinated biphenyls (DL-PCBs)], and chlorinated dibenzo-*p*-dioxins and polychlorinated dibenzofurans (PCDD/PCDFs or PCDD/Fs) were profiled in all matrices examined. Furthermore, the primary method for determining PBDEs is gas chromatography-high-resolution mass spectrometry (GC-HR-MS), a method used to quantify PCDD/PCDF when the highest sensitivity requirements are imposed [71].

Chlorinated paraffins (CPs) exist as congeners in groups (short, medium, and long carbon chain lengths), and are mainly manufactured for industrial use. Through the symbiotic need for the production and utilization of products containing these complex compounds, humans are indirectly exposed to their toxic effects. Of significant exposure is by food intake. Thus, contamination levels of short and medium carbon chain CP (SCCP and MCCP) in meat and meat products were investigated using 2-dimensional GC paired with electron capture negative ionization (ECNI) HR-ToFMS [47]. In the meat samples evaluated, forty-eight homolog groups of CPs were screened. A similar study using GC connected with ECNI HR-Orbitrap mass analyzer was conducted to detect SCCPs and MCCPs in salmon fish samples [46]. Despite spectrometric interferences of other complex compound masses, the analytical technique employed gave a consistent CP homolog pattern. Considering the above studies of samples from animal origin, GC-HRMS could be suggested as an excellent screening method of CPs, irrespective of compositing mass interferences.

Toxophane is a class of POPs with potential toxicity/carcinogenic effect in the human system through ingestion of foods derived mostly from animal origin [72]. Bearing this in mind, Jiang et al. [61] investigated three indicative congeners of the chemical pollutant level in commonly consumed foods by isotope dilution high-resolution gas chromatography/high resolution mass spectrometry (ID-GC-HRMS). Elucidation of the contaminant level was achieved at sub-picogram amount (0.67–12.87

pg/g ww), much owing to the high screening capabilities of the technique employed.

### 8.3.3   OTHER COMPOUNDS

The presence of major medicinal components in plants is known, however, other constituents may cause adverse effects, thus profiling those undesirable metabolites with powerful analytical tools is paramount. Wang and co-workers [66] examined the application of high-resolution GC-MS using liquid-liquid extraction method with selected ion detector, to characterize and quantitatively analyzed bioactive alkylphenols present in *Ginkgo biloba* (ginkgo or gingko). Individual isomers of ginkgolic acids showed 0.5 and 1.5 ppm detection and quantitation limits, respectively, while monitoring at reference ion of $m/z$ 161. Aryl-polysiloxane HP-88 capillary GC column gave a distinct resolution of the positional double bond isomers of ginkgolic acids, which were found unquantifiable in earlier investigations via GC-MS techniques [73, 74]. Furthermore, the high resolution GC-MS quantitation data obtained compared favorably with an accepted high-performance liquid chromatography method. Plastina et al. [62] applied HRGC for the comparative investigation of fatty acid profiles of two varieties of rice. Distinct differences in the individual fatty acids content of rice obtained from diverse geographical regions were established. Hence, accurate nutritional information could be enhanced by HRMS profiling. Oligosaccharides are a form of carbohydrates made up of higher degrees of polymerization (DP) with prebiotic effects [75]. They are known to be comprised of heavy polymers (DP>3), which may require a chemical degradation process that would consequently bring about deviation from true measurements. As a result, accurate analytical determination of their presence in foods is of great importance. In an attempt to detect the content and composition of higher DP oligosaccharides containing foods, while omitting hydrolysis process, prompted an alternative use of low temperature GC-HRMS technique by Montilla et al. [64]. Interestingly, their report indicated oligosaccharides with a DP of up to 7, distinct carbohydrates spectra, carbohydrate level present were consistent and equally agreed well with manufacturer's label. Additionally, the relatively cheap and flexible nature of the method infers a better laboratory tool for screening carbohydrates-containing foods.

Most often, the conversion of raw food to processed products is commonly accompanied by the formation of desirable or offensive sensations as well as modifications in inherent compounds [76]. This therefore necessitates the monitoring of changes in food components, relatively induced by choice of processing method, processing conditions, nature of food constituents among other several factors. Sequel to the foregoing, Giri et al. [65] explored high-resolution olfactometry to monitor the influence of cooking conditions and methods on meat odorants. Using PCA to reduce sample variability effects, the multidimensional GC technique revealed 68 odor-active compounds, while unmasking co-elution of optimum odor zones. With this development, consensual aroma profiling of complex food matrix interferences implies real-time monitoring effects on food safety.

## 8.4   CONCLUSION AND FUTURE PROSPECTS

Despite numerous studies that position GC-HRMS as a highly effective and analytical golden tool for the quantitation and full-scan identification of targeted and untargeted compounds, the occurrence of unanticipated analyte effects or poor optimization of the analytical parameter could result in inaccurate results. GC is also limited to volatile samples and non-polar compounds. While derivatization has largely assisted in this regard [17], limitations still exist regarding the type and form of analytes that can be investigated using this platform.

Despite these drawbacks, the current and future applications of this analytical platform cannot be overemphasized, as described in the highlighted studies reporting of its use. While this will continue, further studies into its use should still be addressed. As highlighted in the previous paragraphs regarding the use of GC-MS for volatile compounds, much more emphasis and analytical studies should be done to develop techniques that would suit analysis of other compounds of interest. An example of such is mycotoxins, which are largely polar compounds, with some studies already indicating the possibility of using GC-MS for its analysis. The precision and sensitivity of high resolution GC-MS further opens huge possibilities for more studies in this regard. Nonetheless, GC-HRMS still remains the method of choice due to its speed, high-resolution capability, allows utilization of mass spectral database, and ease of use. In addition to known metabolites, new knowns and unknowns can be detected in samples of interest and for that purpose, high resolution mass spectrometry (HRMS) would continue to be an essential tool for the identification of such compounds.

## ACKNOWLEDGMENTS

Supported from the National Research Foundation (NRF) of South Africa Scarce Skills Fellowship (Grant no: 120751), University of Johannesburg Global Excellence and Stature (GES) 4.0 Catalytic Initiative Grant, NRF Thuthuka funding (Grant no: 121826), NRF National Equipment Program (Grant no: 99047) and University Research Committee Grant are duly acknowledged.

## KEYWORDS

- **collisionally activated dissociation**
- **degrees of polymerization**
- **dioxin-like polychlorinated biphenyls**
- **food analysis**
- **gas chromatography**
- **high resolution mass spectrometry**

## REFERENCES

1. Lehotay, S. J., & Hajšlová, J., (2002). Application of gas chromatography in food analysis. *Trends Anal. Chem., 21*(9, 10), 686–697.
2. Kachlicki, P., Piasecka, A., Stobiecki, M., & Marczak, L., (2016). Structural characterization of flavonoid glycoconjugates and their derivatives with mass spectrometric techniques. *Molecules, 21*(11), 1494–1514.
3. Adebo, O. A., Kayitesi, E., Tugizimana, F., & Njobeh, P. B., (2019). Differential metabolic signatures in naturally and lactic acid bacteria (LAB) fermented ting (a Southern African food) with different tannin content, as revealed by gas chromatography mass spectrometry (GC–MS)-based metabolomics. *Food Res. Int., 121*, 326–335.

4. Koek, M. M., Jelleme, R. H., Van, D. G. J., Tas, A. C., & Hankemeier, T., (2011). Quantitative metabolomics based on gas chromatography mass spectrometry: Status and perspectives. *Metabolomics, 7*(3), 307–328.

5. Qiu, Y., & Reed, D., (2014). Gas chromatography in metabolomics study. In: Guo, X., (ed.), *Advances in Gas Chromatography* (pp. 83–101). In Tech: Croatia.

6. Garcia, A., & Barbas, C., (2011). Gas chromatography-mass spectrometry (GC-MS)-based metabolomics. In: Metz, D. O., (ed.), *Metabolic Profiling, Methods in Molecular Biology* (pp. 191–204). Springer: Science+Business Media, New York.

7. Čajka, T., & Hajšlová, J., (2004). Gas chromatography-high-resolution time-of-flight mass spectrometry in pesticide residue analysis: Advantages and limitations. *J. Chromatogr. A., 1058*(1/2), 251–261.

8. Brits, M., Gorst-Allman, P., Rohwer, E. R., De Vos, J., De Boer, J., & Weiss, J. M., (2018). Comprehensive two-dimensional gas chromatography coupled to high-resolution time-of-flight mass spectrometry for screening of organohalogenated compounds in cat hair. *J. Chromatogr. A, 1536*, 151–162.

9. Uclés, S., Uclés, A., Lozano, A., Martínez, B. M. J., & Fernández-Alba, A. R., (2017). Shifting the paradigm in gas chromatography mass spectrometry pesticide analysis using high-resolution accurate mass spectrometry. *J. Chromatogr. A, 1501*, 107–116.

10. Pinu, F. R., & Villas-Boas, S. G. F., (2017). Rapid quantification of major volatile metabolites in fermented food and beverages using gas chromatography-mass spectrometry. *Metabolites, 7*(3), 1–13.

11. Wang, S. W., Hsu, K. H., Huang, S. C., Tseng, S. H., Wang, D. Y., & Cheng, H. F., (2019). Determination of polycyclic aromatic hydrocarbons (PAHs) in cosmetic products by gas chromatography-tandem mass spectrometry. *J. Food Drug Anal., 27*(3), 815–824.

12. Mol, H. G. J., Tienstra, M., & Zomer, P., (2016). Evaluation of gas chromatography-electron ionization-full scans high-resolution orbitrap mass spectrometry for pesticide residue analysis. *Anal. Chim. Acta, 935*, 161–172.

13. Laganá, A., & Cavaliere, C., (2015). High-resolution mass spectrometry in food and environmental analysis. *Anal. Bioanal. Chem., 407*, 6235–6236.

14. Kyle, P. B., (2017). Toxicology: GCMS. In: Nair, H., & Clarke, W., (eds.), *Mass Spectrometry for the Clinical Laboratory* (pp. 131–163). Elsevier: Netherlands.

15. Kim, I. Y., Suh, S. H., Lee, I. K., & Wolfe, R. R., (2016). Applications of stable, nonradioactive isotope tracers in *in-vivo* human metabolic research. *Experimental and Molecular Medicine, 48*, e203.

16. Arrebola-Liébanas, F. J., Romero-González, R., & Frenich, A. G., (2017). HRMS: Fundamentals and basic concepts. In: Romero-González, R., & Frenich, A. G., (eds.), *Application in High Resolution Mass Spectrometry; Food Safety and Pesticide Residue Analysis* (pp. 1–14). Elsevier, Amsterdam AE/Oxford, UK/Cambridge, MA, US.

17. Adebo, O. A., (2019). *Metabolomics, Physicochemical Properties and Mycotoxin Reduction of Whole Grain Ting (a Southern African Fermented Food) Produced via Natural and Lactic Acid Bacteria (LAB) Fermentation*. DTech Thesis, Faculty of Science, University of Johannesburg, Johannesburg, South Africa.

18. Harvey, D. J., (2019). Mass spectrometry: Ionization methods overview. In: Worsfold, P., Alan, T. A., Poole, C., & Miró, M., (eds.), *Encyclopedia of Analytical Science* (pp. 398–410). Elsevier: US.

19. Melon, F. A., (2003). Mass spectrometry: Principles and instrumentation. In: Caballero, B., Finglas, P., & Toldra, F., (eds.), *Encyclopedia of Food Sciences and Nutrition* (pp. 3739–3749). Academic Press: US.

20. Rockwood, A. L., Kushnir, M. M., & Clarke, N. J., (2018). Mass spectrometry. In: Rifai, N., Horvath, A. R., & Wittwe, C. T., (eds.), *Principles and Application of Clinical Mass Spectrometry* (pp. 33–65). Elsevier: Amsterdam AE/Oxford, UK/ Cambridge, MA, US.

21. Siuzdak, G., (2005). An introduction to mass spectrometry ionization: An excerpt from the expanding role of mass spectrometry in biotechnology (2nd edn.). MCC Press: San Diego. *J. Assoc. Lab. Autom., 9*(2), 50–63.

22. Schäfer, M., (2019). Mass spectrometry: Fundamentals and instrumentation. In:Worsfold, P., Alan, T. A., Poole, C., & Miró, M., (eds.), *Encyclopedia of Analytical Science* (pp. 358–365). Elsevier: US.

23. Horning, E. C., Horning, M. G., Carroll, D. I., Dzidic, I., & Stillwell, R. N., (1973). New picogram detection system based on a mass spectrometer with an external ionization source at atmospheric pressure. *Analytical Chemistry, 45*, 936–943.

24. Standing, K. G., & Ens, W., (2017). Time of flight mass spectrometers. In: Lindon, J. C., Holmes, J. L., & Tranter, G. E., (eds.), *Encyclopedia of Spectroscopy and Spectrometry* (pp. 458–462). Academic Press: US.

25. García-Reyes, J. F., Moreno-González, D., Nortes-Méndez, R., Gilbert-López, B., & Díaz, A. M., (2017). HRMS: Hardware and software. In: Romero-González, R., & Frenich, A. G., (eds.), *Application in High Resolution Mass Spectrometry: Food Safety and Pesticide Residue Analysis* (pp. 15–57). Elsevier: Amsterdam AE/Oxford, UK/Cambridge, MA, US.

26. Stewart, D., Dhungana, S., Calrk, R., Pathmasiri, W., McRitchie, S., & Summer, S., (2015). Omics technologies used in systems biology. In: Fry, R. C., (ed.), *Systems Biology in Toxicology and Environmental Health* (pp. 57–83). Elsevier: Netherlands.

27. Perry, R. H., Cooks, R. G., & Noll, R. J., (2008). Orbitrap mass spectrometry: Instrumentation, ion motion and applications. *Mass Spectrom. Rev., 27*(6), 661–699.

28. Wood, M., (2019). High-resolution mass spectrometry: An emerging analytical method for drug testing. In: Dasgupta, A., (ed.), *Critical Issues in Alcohol and Drugs of Abuse Testing* (pp. 173–188). Elsevier: Netherlands.

29. Špánik, I., & Machyňáková, A., (2017). Recent applications of gas chromatography with high-resolution mass spectrometry. *Journal of Separation Science, 41*(1), 16–179.

30. Chindarkar, N. S., Park, H. D., Stone, J. A., & Fitzgerald, R. L., (2015). Comparison of different time of flight-mass spectrometry modes for small molecule quantitative analysis. *J. Anal. Toxicol., 39*(9), 675–685.

31. Lin, L., Lin, H., Zhang, M., Dong, X., Yin, X., Qu, C., & Ni, J., (2015). Types, principle, and characteristics of tandem high-resolution mass spectrometry and its applications. *RSC Adv., 5*(130), 107623–107636.

32. Makarov, A., (2019). Orbitrap journey: Taming the ion rings. *Nat. Commun., 10*(1), 3743.

33. Rajawat, J., & Jhingan, G., (2019). Mass spectrometry. In: Misra, G., (ed.), *Data Processing Handbook for Complex Biological Data Sources* (pp. 1–20). Elsevier: Netherlands.
34. Alvarez-Rivera, G., Ballesteros-Vivas, D., Parada-Alfonso, F., Ibanez, E., & Cifuentes, A., (2019). Recent applications of high-resolution mass spectrometry for the characterization of plant natural products. *TRAC, 112*, 87–101.
35. Hammann, S., Korf, A., Bull, I. D., Hayen, H., & Cramp, L. J. E., (2019). Lipid profiling and analytical discrimination of seven cereals using high temperature gas chromatography coupled to high-resolution quadrupole time-of-flight mass spectrometry. *Food Chem., 282*, 27–35.
36. Elss, S., Preston, C., Appel, M., Heckel, F., & Schreier, P., (2006). Influence of technological processing on apple aroma analyzed by high-resolution gas chromatography-mass spectrometry and on-line gas chromatography-combustion/pyrolysis-isotope ratio mass spectrometry. *Food Chem., 98*(2), 269–276.
37. Belz, M. C. E., Axel, C., Beauchamp, J., Zannini, E., Arendt, E. K., & Czerny, M., (2017). Sodium chloride and its influence on the aroma profile of yeasted bread. *Foods, 6*(8), 66.
38. Duan, Y., Zheng, F., Chen, H., Huang, M., Xie, J., Chen, F., & Sun, B., (2015). Analysis of volatiles in Dezhou braised chicken by comprehensive two-dimensional gas chromatography/high resolution-time of flight mass spectrometry. *LWT-J. Food Sci. Technol., 60*(2), 1235–1242.
39. Wang, W., Feng, X., Zhang, D., Li, B., Sun, B., Tian, H., & Liu, Y., (2018). Analysis of volatile compounds in Chinese dry-cured hams by comprehensive two-dimensional gas chromatography with high-resolution time-of-flight mass spectrometry. *Meat Sci., 140*, 14–25.
40. Bader, S., Czerny, M., Eisner, P., & Buettner, A., (2009). Characterization of odor-active compounds in lupin flour. *J. Sci. Food Agr., 89*(14), 2421–2427.
41. Gracka, A., Majcher, M., Kludská, E., Hradecký, J., Hajšlová, J., & Jelen, H. H., (2018). Storage-induced changes in volatile compounds in argan oils obtained from raw and roasted kernels. *J. Am. Oil Chem. Soc., 95*(12), 1475–1485.
42. Janzantti, N. S., Macoris, M. S., Garruti, D. S., & Monteiro, M., (2012). Influence of the cultivation system in the aroma of the volatile compounds and total antioxidant activity of passion fruit. *LWT-J. Food Sci. Technol., 46*(2), 511–518.
43. Cossignani, L., Urbani, E., Simonetti, M. S., Maurizi, A., Chiesi, C., & Blasi, F., (2014). Characterization of secondary metabolites in saffron from central Italy (Cascia, Umbria). *Food Chem., 143*, 446–451.
44. Stupak, M., Goodall, I., Tomaniova, M., Pulkrabova, J., & Hajslova, J., (2018). A novel approach to assess the quality and authenticity of scotch whisky based on gas chromatography coupled to high-resolution mass spectrometry. *Anal. Chim. Acta, 1042*, 60–70.
45. Dunovska, L., Cajka, T., Hajslova, J., & Holadiva, K., (2006). Direct determination of acrylamide in food by gas chromatography-high-resolution time-of-flight mass spectrometry. *Anal. Chim. Acta, 578*(2), 234–240.
46. Krätschmer, K., Cojocariu, C., Schächtele, A., Malisch, R., & Vetter, W., (2018). Chlorinated paraffin analysis by gas chromatography orbitrap high-resolution mass

spectrometry: Method performance, investigation of possible interferences and analysis of fish samples.*J. Chromatogr. A, 1539,* 53–61.

47. Huang, H., Gao, L., Zheng, M., Li, J., Zhang, L., Wu, Y., Wang, R., et al., (2018). Dietary exposure to short- and medium-chain chlorinated paraffins in meat and meat products from 20 provinces of China. *Environ. Pollut., 233,* 439–445.

48. Schallschmidt, K., Hitzel, A., Po¨hlmann, M., Schwa¨gele, F., Speer, K., & Jira, W., (2012). Determination of 3-MCPD in grilled meat using pressurized liquid extraction and gas chromatography-high resolution mass spectrometry. *J. Consum. Prot. Food S., 7*(3), 203–310.

49. Godliauskienė, R., Tamošiūnas, V., & Naujalis, E., (2017). Polychlorinated dibenzo-p-dioxins/furans and dioxin-like polychlorinated biphenyls in food and feed in the Lithuanian market. *Environ. Toxicol. Chem., 99*(1), 65–77.

50. Rjabova, J., Viksna, A., & Zacs, D., (2018). Development and optimization of gas chromatography coupled to high-resolution mass spectrometry-based method for the sensitive determination of dechlorane plus and related norbornene-based flame-retardants in food of animal origin. *Chemosphere, 191,* 597–606.

51. Zacs, D., Perkons, I., Volkovs, V., & Bartkevics, V., (2019). Multi-analyte method for the analysis of legacy and alternative brominated and chlorinated flame-retardants in food products of animal origin using gas chromatography-magnetic sector high-resolution mass spectrometry. *Chemosphere, 230,* 396–405.

52. Rozentale, I., Zacs, D., Perkons, I., & Bartkevics, V. A., (2017). Comparison of gas chromatography coupled to tandem quadrupole mass spectrometry and high-resolution sector mass spectrometry for sensitive determination of polycyclic aromatic hydrocarbons (PAHs) in cereal products. *Food Chem., 221,* 1291–1297.

53. Martorell, I., Perelló, G., Martí-Cid, R., Castell, V., Llobet, J. M., & Domingo, J. L., (2010). Polycyclic aromatic hydrocarbons (PAH) in foods and estimated PAH intake by the population of Catalonia, Spain: Temporal trend. *Environ. Int., 36*(5), 424–432.

54. Jira, W., Ziegenhals, K., & Speer, K., (2008). Gas chromatography-mass spectrometry (GC-MS) method for the determination of 16 European priority polycyclic aromatic hydrocarbons in smoked meat products and edible oils. *Food Addit. Contam.A, 25*(6), 704–713.

55. Wretling, S., Eriksson, A., Eskhult, G. A., & Larsson, B., (2010). Polycyclic aromatic hydrocarbons (PAHs) in Swedish smoked meat and fish. *J. Food Compos. Anal., 23*(3), 264–272.

56. Lopez-Ruiz, R., Romero-Gonzalez, R., Serra, B., & Frenich, A. G., (2019). Dissipation kinetic studies of fenamidone and propamocarb in vegetables under greenhouse conditions using liquid and gas chromatography coupled to high-resolution mass spectrometry. *Chemosphere, 226,* 36–46.

57. Pruteanu, E., Niculita, P., Catană, L., Catană, M., Iorga, M. N. E., & Bălea, A., (2013). Dioxins and furans eggs contamination evaluated using high-resolution gas chromatography coupled to high-resolution mass spectrometry. *J. Agroaliment. Processes Technol., 19*(1), 83–87.

58. Ten, D. G., Pussente, I. C., Scholl, G., Eppe, G., Schaechtele, A., & Van, L. S., (2016). The performance of atmospheric pressure gas chromatography-tandem mass spectrometry compared to gas chromatography-high resolution mass spectrometry

for the analysis of polychlorinated dioxins and polychlorinated biphenyls in food and feed samples. *J. Chromatogr. A, 1477*, 76–90.

59. Zhang, L., Yin, S., Wang, X., Li, J., Zhao, Y., Li, X., Shen, H., & Wu, Y., (2015). Assessment of dietary intake of polychlorinated dibenzo-p-dioxins and dibenzofurans and dioxin-like polychlorinated biphenyls from the Chinese total diet study in 2011. *Chemosphere, 137*, 178–184.

60. Čonka, K., Fabišiková, A., Chovancová, J., Sejáková, Z. S., Dömötörová, M., Drobná, B., & Kočan, A., (2015). Polychlorinated dibenzo-p-dioxins, dibenzofurans and biphenyls in food samples from areas with potential sources of contamination in Slovakia. *J. Food Nutr. Res., 54*(1), 50–61.

61. Jiang, Y., Liu, Z., Wu, D., Zhang, J., Zhou, J., Li, S., Lu, L., et al., (2016). Toxaphene levels in retail food from the Pearl River Delta area of South China and an assessment of dietary intake. *Chemosphere, 152*, 318–327.

62. Plastina, P., Gabriele, B., & Fazio, A., (2018). Characterizing traditional rice varieties grown in temperate regions of Italy: Free and bound phenolic and lipid compounds and *in vitro* antioxidant properties. *J. Food Saf. Food Qual., 2*(2), 89–95.

63. Li, X., Lv, P., Wang, L., Guo, A., Ma, M., & Qi, X., (2014). Application of high-resolution pyrolysis gas chromatography/mass spectrometry (HRPGC/MS) for detecting listeria monocytogenes. *J. Chromatogr. B, 971*, 107–111.

64. Montilla, A., Van, D. L. J., Olano, M., & Del, C. M. D., (2006). Determination of oligosaccharides by conventional high-resolution gas chromatography. *Chromatographia, 63*(9/10), 453–458.

65. Giri, A., Khummueng, W., Mercier, F., Kondjoyan, N., Tournayre, P., Meurillon, M., Ratel, J., & Engel, E., (2015). Relevance of two-dimensional gas chromatography and high-resolution olfactometry for the parallel determination of heat-induced toxicants and odorants in cooked food. *J. Chromatogr. A, 1388*, 217–226.

66. Wang, M., Zhao, J., Avula, B., Wang, Y. H., Avonto, C., Chittiboyina, A. G., Wylie, P. L., et al., (2014). High-resolution gas chromatography/mass spectrometry method for characterization and quantitative analysis of ginkgolic acids in ginkgo biloba plants, extracts, and dietary supplements. *J. Agric. Food Chem., 62*(50), 12103–12111.

67. Yan, D., Wong, Y. F., Shellie, R. A., Marriot, P. J., Whittock, S. P., & Koutoulis, A., (2019). Assessment of the phytochemical profiles of novel hop (*Humulus lupulus* L.) cultivars: A potential route to beer crafting. *Food Chem., 275*, 15–23.

68. Romera-Torres, A., Arrebola-Liebanas, J., Vidal, J. L. M., & Frenich, A. G., (2019). Determination of calystegines in several tomato varieties based on GC-Q-Orbitrap analysis and their classification by ANOVA. *J. Agr. Food Chem., 67*(4), 1284–1291.

69. Adebo, O. A., Kayitesi, E., Adebiyi, J. A., Gbashi, S., Phoku, J. Z., Lasekan, A., & Njobeh, P. B., (2017). Mitigation of acrylamide in foods: An Africa perspective. In: Boreddy, S. R. R., (ed.), *Acrylic Polymers in Healthcare* (pp. 151–174). InTech: Croatia.

70. WHO, World Health Organization, (2016). *Dioxins and Their Effect on Human Health*. World Health Organization Media Centre.

71. Shelepchikov, A. A., Ovcharenko, V. V., Kozhushkevich, A. I., Brodskii, E. S., Komarov, A. A., Turbabina, K. A., & Kalantaenko, A. M., (2019). A new method for purifying fat containing extracts in the determination of polybrominated diphenyl ethers. *J. Anal. Chem., 74*(6), 574–558.

72. Buranatrevedh, S., (2004). Cancer risk assessment of toxaphene. *Ind. Health, 42*(3), 321–327.
73. Sun, Y., Tang, C., Wu, X., Pan, Z., & Wang, L., (2012). Characterization of alkylphenol components in ginkgo biloba sarcotesta by thermochemolysis-gas chromatography/mass spectrometry in the presence of trimethylsulfonium hydroxide. *Chromatographia, 75*(7/8), 387–395.
74. Wang, L., Jia, Y., Pan, Z., Mo, W., & Hu, B., (2009). Direct analysis of alkylphenols in ginkgo biloba leaves by thermochemolysis-gas chromatography/mass spectrometry in the presence of tetramethylammonium hydroxide. *J. Anal. Appl. Pyrol., 85*(1/2), 66–71.
75. Tymczyszyn, E. E., Santos, M. I., Costa, M. C., Illanes, A., & Gómez-Zavaglia, A., (2013). History, synthesis, properties, applications, and regulatory issues of prebiotic oligosaccharides. In: Gil, M. H., (ed.), *Carbohydrates Applications in Medicine* (pp. 1–28). Research Signpost, India.
76. Adebo, O. A., & Medina-Meza, I. G., (2020). Impact of fermentation on the phenolic compounds and antioxidant activity of whole cereal grains: A mini review. *Molecules, 25*(4), 927.

# Index

Printed and bound by CPI Group (UK) Ltd, Croydon, CR0 4YY

23/10/2024

01777702-0004